SOIL EROSION IN INDIA

SOIL EROSION IN INDIA

E. AHMAD

Professor and
Head of the Department of Geography,
Ranchi University, Ranchi

ASIA PUBLISHING HOUSE

LONDON

Enayat **Ahmad** (*b.* 1923)

ISBN 0.210.22247.6

PRINTED IN INDIA

BY PARIMAL BOSE AT THE BASUSREE PRESS, 80/6, GREY STREET, CALCUTTA AND PUBLISHED BY P. S. JAYASINGHE, ASIA PUBLISHING HOUSE, 447 STRAND, LONDON. W.C. 2

PREFACE

ONE of the common defects of books on soil erosion and conservation in India is the absence of an all-India picture in which a geographer in particular and the average reader in general is keenly interested. Indeed, the local details are essential but we are impatient to know where the details stand in the larger context. The importance of neither the details nor the larger framework can be minimised, but the latter has probably been invariably overlooked. This little book undertakes to pay particular attention to the larger framework of soil erosion in India. The different types of soil erosion in the country have so far probably not been discussed with reference to their regional and relative magnitude, much less with reference to their regional causes—biotic, meteorological, tectonic, etc. An areal estimate of erosion within the different districts has not been attempted so far. This book tries to do such things. It must be confessed that it is not a technical work. It is meant for the largest number of readers. Yet technical groups, *e.g.* planners, geographers, soil scientists, agriculturists and geologists, are likely to benefit from its study.

Ranchi

December 1967 ENAYAT AHMAD

CONTENTS

MAPS AND DIAGRAMS

CHAPTER I

INTRODUCTION

Every parcel of the ground that is covered with soil and is exposed to weathering and erosion is subject to soil erosion of one kind and degree or the other. It is plain that all the surface of India, except the completely rocky soil-less surface on the steep slopes of mountains, hills and plateau edges, or the permanently ice-covered Himalayan region, or water bodies, or built-up areas, or other small tracts physically protected against erosion, undergoes soil erosion. In recent years, particularly after Independence, there has been a growing consciousness about soil erosion in India.

Definite soil conservation schemes are in progress in the areas of spectacular erosion (Fig. 14), particularly in the ambit of the major river valley projects, *e.g.* Hazaribagh district in the D.V.C., Bilaspur district in the catchment of Bhakra-Nangal Project and Koraput district in the catchment of Machkund Project etc. Probably detailed work is being done in the districts under soil conservation schemes and useful information and maps may be produced in due course. The layman, however, may be impatient to know the national or regional magnitude of the malady of soil erosion. What, generally speaking, are the factors that cause, control or influence soil erosion in different parts of India? What are the processes of erosion in the Indian scene? What is the distribution of the different types of erosion and what are the problem areas in the country? Is it possible to estimate the quantity of soil that is eroded every year, or to estimate the area occupied by the different types of soil erosion? What is the rate, in general terms, at which erosion has been increasing? Can the amount of agricultural production lost by soil erosion be calculated and so on? What are the conservation measures suited to the different parts of the country?

There is a vast treasure of literature* on soil erosion and conservation published in different countries of the world and recently in India. A good deal of the Indian literature is theoretical or is related to regional and local field studies. Any amount of theoretical knowledge cannot help us in finding precise answers to the questions posed above. On the other hand, it is possible neither now nor in future to carry out field work in every plot of land or even in every village. On the basis of micro-regional sample observations it might be possible in future to obtain relatively precise answers to these questions and to prepare detailed soil erosion maps. The author guesses that such a time may be two to three decades hence. In the meantime, something can be done to inform the average reader about the magnitude and nature of the problem.

The basic facts of geology, the force and regime of rainfall, wind action, geomorphology, hydrology, vegetation cover, the common properties of soils, crop pattern, the biotic and other factors of the environment whose integrated knowledge is the special equipment of the geographer can help us to some extent and very broadly, of course, in answering at least some of the questions posed above. Some such answers, howsoever vague and generalized, are the scope of this book.

* To mention a few, we have the following :

(i) Bennett, Hugh Hammond, 'Soil Conservation' (McGraw-Hill, New York, 1939).

(ii) Publications of the U.S. *Soil Conservation Service.*

(iii) Gustafson, 'Soil and Soil Management' (McGraw-Hill, New York, 1941).

(iv) The Yearbook of Agriculture (U.S. Dept. of Agriculture), (Water), 1955, (Soil), 1957.

(v) 'Bhagirath'—The Irrigation and Power Monthly (Publications Division, Delhi—Vol. IX, No. 3 (1962), Vol. II, No. 8 (1956), Vol. VII, No. 4 (1961).

(vi) 'Journal of Soil and Water Conservation in India' (Hazaribagh, India)—since 1952.

(vii) 'Soil Conservation Programme for Agricultural Land'; Programme Evaluation Organization Publication, No. 41, 1964 (Delhi, 1965).

(viii) Rao, M. S. V., 'Soil Conservation in India' (ICAR, New Delhi, 1962).

Physical basis of Soil erosion: It is fairly certain that the degree and nature of soil erosion varies with the physical factors enumerated above. It is possible to divide India into regions in which the nature and extent of soil erosion are largely similar. Thus, the largest units or the regions of the first order are: (1) The Himalaya and adjoining mountains and hills, (2) Indo-Gangetic Plain, (3) the Rajasthan desert and its fringes, (4) the Peninsular Upland and (5) the Coastal Plains. These first-order regions have physical variations on the basis of which the following second-order regions in respect of soil erosion may be delimited:

The Himalaya & adjoining highlands.

(1) (i) The snow-covered Great Himalaya.
(1) (ii) The Lesser Himalaya.
(1) (iii) The 'Dun', or tectonic valleys.
(1) (iv) The foothills, or Siwaliks.
(1) (v) The North-eastern hills.

The Indo-Gangetic Plain.

(2) (i) The *Cho* & *Bhabar* zones.
(2) (ii) The *Tarai*.
(2) (iii) The Indus basin.
(2) (iv) The Indo-Gangetic divide.
(2) (v) The Ganga Plain.
(2) (vi) The Assam Valley.

The Rajasthan desert and its fringes.

(3) (i) The 'Thar' or the desert proper.
(3) (ii) The fringes of the desert.

The Peninsular Upland.

(4) (i) The Central Indian Foreland.
(4) (ii) North-eastern uplands.
(4) (iii) The lava region.
(4) (iv) Deccan Proper.

The Coastal Plains.

(5) (i) The Western Coastal Plain.
(5) (ii) The Eastern Coastal Plain south of the Godavari delta.
(5) (iii) The Eastern Coastal Plain north-east of the Godavari delta.

Again, within these second-order regions, there are considerable variations and even a cursory study of the physical basis of soil erosion in India must take them into consideration. Thus, we have the third-order regions as follows (Fig. 1):

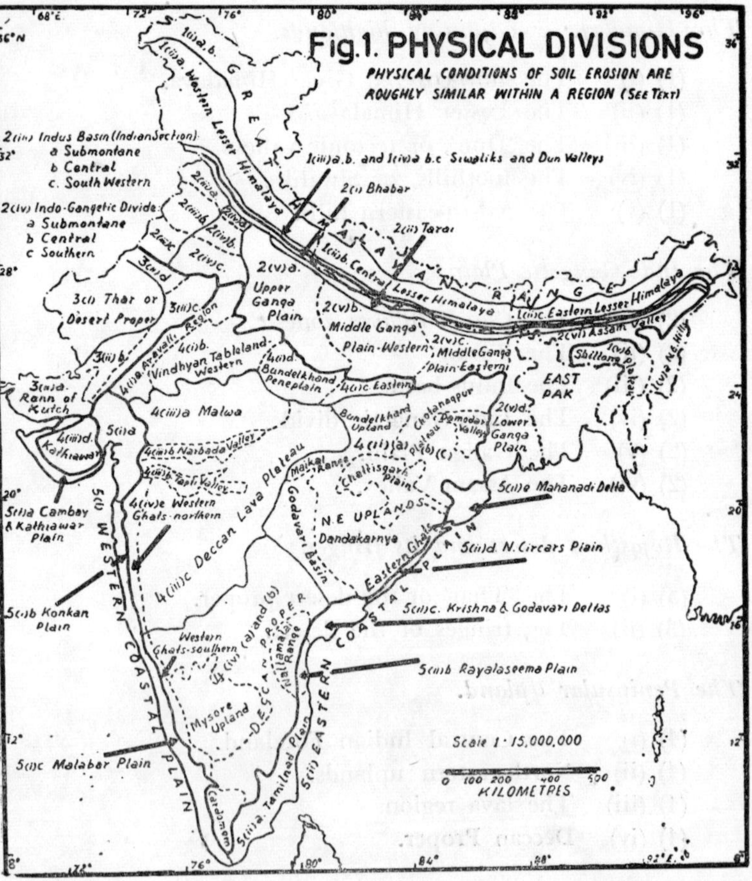

Fig. 1. PHYSICAL DIVISIONS

The Himalaya and adjoining highlands.

(1) (i) (a)	Permanently snow-covered Great Himalaya.
(1) (i) (b)	The Periglacial or lower and marginal sections near the border of the Great and Lesser Himalaya, subject to summer thaw or heavy winter snow.
(1) (ii) (a)	The Western Lesser Himalaya.
(1) (ii) (b)	The Central Lesser Himalaya.
(1) (ii) (c)	The Eastern Lesser Himalaya.
(1) (iii) (a)	The Western tectonic valleys.
(1) (iii) (b)	The Central tectonic valleys.
(1) (iv) (a)	The Western foothills.
(1) (iv) (b)	The Central foothills.
(1) (iv) (c)	The Eastern foothills.
(1) (v) (a)	The North-eastern hills.
(1) (v) (b)	Shillong plateau.

The Indo-Gangetic Plain.

(2) (i)	The *Bhabar* region.
(2) (ii)	The *Tarai.*
(2) (iii) (a)	The submontane Indus basin.
(2) (iii) (b)	The central Indus basin (Indian section).
(2) (iii) (c)	The south-western Indus basin (Indian section).
(2) (iv) (a)	The Indo-Gangetic divide—submontane.
(2) (iv) (b)	„ „ „ „ —central.
(2) (iv) (c)	„ „ „ „ —south-western

The Ganga Plain.

(2) (v) (a)	The Upper Ganga Plain.
(2) (v) (b)	The Western middle Ganga Plain.
(2) (v) (c)	The Eastern middle Ganga Plain.
(2) (v) (d)	The Lower Ganga Plain.
(2) (vi)	Assam Valley.

The Indian desert and its fringes.

(3) (i)	The *Thar,* or desert proper.
(3) (ii) (a)	Rann of Kutch.
(3) (ii) (b)	The Luni basin.
(3) (ii) (c)	The eastern fringes of the desert.
(3) (ii) (d)	The northern fringes of the desert.

The Peninsular Upland.

(4) (i) (a)	The Aravalli region.
(4) (i) (b)	The Vindhyan tableland–western.
(4) (i) (c)	The Vindhyan tableland–eastern, and adjoining areas.
(4) (i) (d)	Bundelkhand gneissic peneplain and adjoining areas.

North-eastern Uplands.

(4) (ii) (a)	Scarps and hills, *e.g.* in Chotanagpur, Orissa, Rajmahals and Baghelkhand.
(4) (ii) (b)	Peneplains, *e.g.* Ranchi and Chaibasa plains.
(4) (ii) (c)	Valleys, *e.g.* the Damodar, Subarnarekha, the Godavari and the Mahanadi.

The lava region.

(4) (iii) (a)	Malwa.
(4) (iii) (b)	The Narmada and Tapti valleys.
(4) (iii) (c)	Deccan lava plateau.
(4) (iii) (d)	Kathiawar.
(4) (iii) (e)	The Western Ghats.

Deccan Proper.

(4) (iv) (a)	Deccan plateaus and peneplains.
(4) (iv) (b)	Deccan scarps and hills.

Coastal Plains.

(5) (i) (a)	Kathiawar and Cambay coastal plains.
(5) (i) (b)	The Konkan coastal plain.
(5) (i) (c)	The Malabar coastal plain.

Eastern Coastal Plain.

(5) (ii) (a) The Tamilnad coastal plain.
(5) (ii) (b) The Rayalaseema coastal plain.
(5) (ii) (c) The Krishna and Godavari deltas.
(5) (ii) (d) The coastal plain of Northern Circars.
(5) (ii) (e) The Mahanadi delta.

CHAPTER II

FACTORS OF SOIL EROSION

THE PHYSICAL conditions of soil erosion in the different regions mentioned in the foregoing chapter (Fig. 1) are roughly uniform within the different third-order regions. But such a regional pattern is over-simplified. The different factors of soil erosion may transcend the physical boundaries laid down in the map. The working of the erosion factors is extremely complicated. It is only in the broadest and simplest outlines that we propose to examine the factors affecting or controlling soil erosion in the different parts of India. In other words, we shall discuss the geographical distribution of soil erosion factors in the country. The chief factors of erosion in India are :

(*i*) Rainfall.
(*ii*) Floods.
(*iii*) Snow.
(*iv*) Wind.
(*v*) Vegetation.
(*vi*) Zoological factor.
(*vii*) Geological factor.
(*viii*) Geomorphic factor.
(*ix*) Sea.

(*i*) RAINFALL

Except the Himalayas in cold weather, precipitation in India is mostly in the form of rainfall which is generally of the nature of forceful downpours characteristic of monsoonal rainfall. The monsoonal showers are particularly forceful towards the beginning and end of the S. W. monsoon period, when the monsoonal depressions assume the nature of tropical cyclones. Even during the hot season (April to mid-June) precipitation brought by duststorms, or thunderstorms is in the form of violent showers.

Winter season is characterised generally by gentle drizzles associated with winter disturbances. Although this rainfall is the maximum in the north-west, most of the country north of the Tropic of Cancer gets some precipitation during the season. South India, particularly the eastern parts, also receives rainfall during the season. It is cyclonic and torrential.

The regime, amount, nature, and distribution of precipitation in India influences soil erosion in the following manner:

The cold season, *i.e.* the period from November to March for the most of the country *north* of the Krishna is a period of minimum soil erosion as a result of low rainfall coming in the shape of drizzles. In eastern Deccan, however, the N. E. monsoon brings torrential rain associated with cyclones, so that this area is subject to marked soil erosion.

During the hot season from April to mid-June, the zones which are most liable to soil erosion by rain are north-east India (Assam, Manipur, Nagaland, Tripura, West Bengal and eastern Bihar) having the greatest frequency of thunderstorms or nor'westers. The intensity of erosion, if any, decreases westward and southward with the decrease of rainfall till it again increases in the extreme south where the pre-monsoon 'mango' showers under the equatorial regime of rainfall are considerable and in the form of tropical downpours.

The main period of soil erosion, however, is the period of south-west monsoon and, neglecting other factors, we may say that the intensity of soil erosion is directly proportionate to the amount of monsoonal rainfall, i.e. the intensity of erosion decreases from the western coastal plain and 'Ghats' towards the rainshadow in Andhra and eastern coastal plain. Again, it rises in the N. E. highlands. In the northern plains and Himalayas, it is expected to decrease from east to west. It is expected to be at its minimum in the Rajasthan desert and at its maximum on the southern slopes of Shillong plateau.

In the above-noted lines, we have concentrated our attention on precipitation (to the exclusion of other factors) as a factor for soil erosion. In India, as elsewhere in the world, rainfall is the chief agent of soil erosion and soil movement. Let us give some thought to the mechanism of soil erosion under rainfall. As rain falls, it moistens the soil into mud or paste or muddy water. Water, therefore, works as a lubricant of soil particles. The greater the amount of rainfall, the greater the depth to which the soil and sub-soil are loosened. Thus, even if there were no slope and the surface were perfectly horizontal guarding against cumulative flow of water, the soil is rendered extremely suitable for erosion by rain whether it is gentle or torrential. The quantity of rainfall and the duration of a shower are important in the sense that the loosening of the soil and the depth to which the loosening takes place are directly proportional to these two factors. The nature of precipitation, however, is more important than mere quantity or duration. Thus, torrential downpour beats upon the soil particles and has a digging effect by itself irrespective of other factors. Torrential downpour is generally characterised by being too much for the area as well as the slope of the land on which it is falling. It cannot be disposed of by soakage or evaporation or gentle flow. It assumes torrential flow which is the most powerful means of soil removal.

Of the precipitation that falls on the ground part is lost by evaporation and part is soaked underground. The balance is available as run-off as a soil-carrying factor. No doubt, the water that is soaked loosens soil particles and this loosening is an important condition for soil erosion, yet the run-off, *i.e.* the part of precipitation that is able to flow out of the point of precipitation is the chief carrier of eroded soil and soil erosion is mainly dependent on the quantity of run-off. The quantitative distribution of run-off in the country is broadly proportionate to the amount of annual rainfall.

(*ii*) Floods

Flood is both soil-depositing and soil-eroding factor. In this connection floods need definition. Floods (Fig. 2) refer to such discharge of water that cannot be contained within the normal channels of drainage. The increased discharge causes a rise of the water surface and the horizontal expansion of water beyond the bank of streams. Settlements, roads, railways and cultivation, etc., are so located that they are safe under the discharge of normal drainage. Under floods, however, they may be invaded by water. Floods normally are caused by concentrated downpours during a relatively short time. They may also result from the damming up of water behind a land-

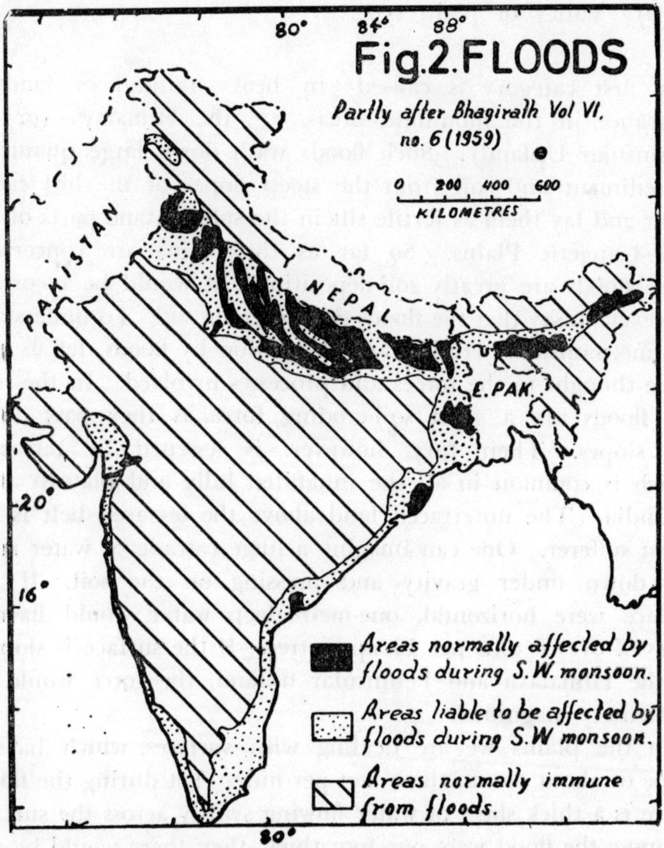

slide in the Himalaya. The obstruction may later be destroyed by the pressure of water causing inundation of the plains below. Floods may also result from the unexpected breach of an artificial reservoir. Apart from such immediate causes, deforestation, the disturbance of drainage lines by such structures as canals, roads and railway embankments, etc., and the changes of level during earthquakes are well known.

For the purposes of soil erosion, floods may be divided into two types :

(*i*) Mountain and hill type.

(*ii*) Valley or plain type.

The first category is caused by heavy rainfall or landslip clearance in the mountain areas, *e.g.* the Himalaya (or the Peninsular Upland). Such floods wash down large quantities of sediment and soil from the steep slopes of the hill catchment and lay them as fertile silt in the submontane parts of the Indo-Gangetic Plains. So far as the plains are concerned, these floods are greatly soil-depositing. It would be incorrect, however, to say that the floods do not erode soil. To understand the mechanism of erosion and deposition by floods, let us give some thought to the forces and processes involved. In the hills the floods are a great soil-eroding force as they pass down the slopes. Their fury, however, is lessened by terracing which is common in all the inhabited hilly and plateau areas of India. The unterraced land above the terraced belt is the worst sufferer. One can imagine a huge cascade of water rushing down under gravity and pressing on the soil. If the surface were horizontal, one-metre-deep water would have a mass of over 6 tons per 10 sq. metres. If the surface is sloping as the Himalaya and Peninsular upland, the force would be immensely magnified.

In the plains we are dealing with surfaces which have a slope of about one to three feet per mile. Yet during the floods there is a thick sheet of water flowing swiftly across the surface. Suppose the flood were one foot thick, then there would be one acre-foot or 43,560 cu. feet of water or about 1,200 tons of water per acre. Such a large mass of water will flow in a short time across a point. It has great eroding force so long as it flows and presses against the soil. In such areas where the flow is retarded in the lee of topographic rises, the settling of silt takes place. Within the streams the flow is strong but beyond the levees or margins, water is separated from the main stream.

This spill has two-directional movement. There is an outflow when the flood is rising and erosion of soil is caused to some extent. Then may follow a period of relative stagnance which will be attended with the maximum settling of silt. This is followed by an inflow towards the stream. This is slow and rarely associated with soil erosion. That even on relatively horizontal surface soil erosion is caused by floods can be seen in the field. Vast expanses of standing crops are uprooted and washed away. The plants are uprooted partly by the pressure of flowing water and partly by the digging effect of water on soil. This, however, is more pronounced where the flow is channellised.

Theoretically, we might say that so long as there is flow in flood erosion is possible. Complete stagnance alone is attended with sedimentation only. Generally speaking, the hill or mountain floods, although damaging to the crops and other human establishments, are a soil-depositing force. Their force may be absorbed within the upper parts of the plains.

The second type of floods may be caused by heavy rainfall within the plains, *e.g.* in Punjab, U.P., Bihar, West Bengal etc. They are not loaded with large quantities of sediments as are the mountain floods. They inundate and wash away crops but their soil-eroding and soil-depositing capacities are limited. Roughly, we may say that floods are largely a soil-eroding force in those parts of India which are 300 metres or about 1,000 feet above the sea level. In the regions between 100 and 300 metres above the sea level, they are mostly soil-eroding and partly soil-depositing while in the country under 100 metres above sea level the floods are mostly a soil-depositing force. Fig. 2 shows the distribution of floods in India. The phenomenon of floods is mostly a feature of the lowlands and plains. In the hill areas, the Himalaya and the Peninsular upland, because of the steep slopes and quick run-off, the discharge rarely grows into floods but the amount of erosion caused by large discharge in these areas of marked and high slopes is colossal.

(*iii*) Snow

Glacial erosion is characteristic of the areas above and near the snow line in the Himalayas. Mostly such areas (Fig. 11) are uninhabited and the effects of such erosion are significant only as eroded sediments passing down through the lower reaches of the streams which originate in or near the snow fields. The periglacial margins may, however, suffer directly by the frequency of avalanches, or streams issuing forth from snow melt. Snow melt gives rise to myriads of tiny streams that run over the ground below. The erosion effect is aggravated by the high gradients that characterise the alluvial cones and fans of mountain periglacial areas. They may bury the alluvial cones that carry the highest cultivation and settlements in the Himalaya near the southern fringe of the Great Himalaya or the transverse openings formed by the major rivers. The advancing avalanches overrun the terraced fields, the terrace walls of which are broken and large amounts of the soils are scrubbed away. Protection is sometimes sought by the planting of line of trees above the arable or putting a wall of the logs of trees. The details of glacial weathering in the high Himalaya do not concern us so far as soil erosion is concerned.

(*iv*) Wind

Wind is one of the factors in soil erosion. The most powerful checks against wind erosion are soil moisture and vegetation cover. Wind erosion of soil is, therefore, confined mainly to the non-monsoon period over north of the country or, more precisely speaking, to the dry period which varies in length from about 3 months in Kerala to practically the whole of the year in the heart of Rajasthan desert. Whereas moisture acting as a thin film of water binds the soil particles, its absence helps to disintegrate them. They are lighter and capable of deflation because of non-cohesion.

Generally speaking, we might say that wind erosion of soil is inversely proportionate to the total amount of rainfall in India. Wind erosion of soil is non-existent in marshy and water-logged regions and as such canals and all other forms of irrigation causing water-logging in the hitherto dry areas are a new environmental factor against soil erosion by wind.

Vegetation in India may be forest, groves, bushes, and grasses, a combination of some or all of these, and cultivated crops which are mostly of the nature of grasses. All these work as checks to soil erosion by wind. The force of wind is reduced by trees and other forms of vegetation. The roots of vegetation, working as binders of soil particles, are checks to soil deflation. Moreover, vegetation, particularly grass and cultivated crops, work as a sort of protective cover or sheet against the detachment and lifting of soil particles by wind.

Now the dry period is associated partly with cultivated crops of winter when the wind is low and sometimes there are light showers also. Thus, during part of the dry season, the soil may be protected against erosion. But the hotter section of the dry period, when humidity is the minimum, is usually a period of cropless landscape over most of India (except the deltas and Malabar Coastal plain etc.). There is no crop protection to soil against wind erosion. Moreover, the deciduous vegetation which is the most extensive and predominant flora in India gets leafless during the hot dry period. Thus, part of the protection afforded by vegetation is gone with the leaves. Leaflessness also increases wind velocity tending to more active soil erosion by wind.

The force of wind is associated with the distribution of air pressure and varies with seasons. Actually, it is the dry period when the wind force is significant in respect to soil erosion. The post-monsoon dry period is immediately followed by a month of slightly high winds when wind erosion may be perceptible in cropless fallow lands especially of light soils. The cold weather dry season is generally a period of gentle winds (3-8 Km. per hour) throughout the country and wind erosion of soil is negligible. The most critical period is the hot weather season when the wind velocity is about 15 to 25 km. per hour. In the northern plains and the desert there is continuous period of strong dusty 'loo' winds from late

March to late May or early June. These winds, reaching their maximum force in the afternoon, are the most effective soil-eroding winds in India. The damage done by them is helped by the coincidence of the 'loo' winds with cropless landscape, minimum atmospheric humidity, leafless vegetation and maximum diurnal range of temperature causing soil cracks and soil disintegration.

An important factor at this time are the summer storms—dust-raising winds, duststorms and thunderstorms—which have their maximum force and frequency during this period. The winds during the storm may reach a gale force of 75 km. per hour or more. These short-lived but strong and violent winds probably effect more soil erosion than the normal winds of the season. The storms have considerably long paths extending over hundreds of kilometres so that the soils picked up from one region may be deflated to a physically different region.

Wind erosion is probably the highest in and around Rajasthan. The adjoining parts, i.e. Punjab, Malwa and Gujarat come next. Next follow the western Ganga Plain and the north-western parts of Peninsular uplands. There is negligible wind erosion of soil in the forested and snow-covered Himalaya, N. E. Mountains, moist Assam Valley and the Ganga delta, the deltas of other major peninsular rivers and wet Malabar Coastal plain. Wind erosion in the other wooded parts of Peninsular India is also restricted, but it may be appreciable in the regions where the annual precipitation may be less than 20 inches as in parts of Andhra and Mysore.

We may now give some thought to the mechanism of wind erosion of soil. While rain acts as a vertical force beating upon the soil from above, wind generally operates as a tangential or translational force transferring the soil particles from one direction to the other. In this sense it is slightly comparable to running water but markedly different from the latter. Running water must take the whole or part of eroded soil as water solution to the local base-level but wind is not capable of such certain action. The larger soil particles are subject to short horizontal and vertical transport only. The larger particles are generally sand or quartz particles and by themselves are relatively infertile and their erosion may be immaterial if their transport does not result in the exposure of soil-less regolith or base rock. The finer soil particles, more productive, are more liable to erosion and this is material to agricultural production. These fine soil particles may be carried aloft and to considerable horizontal distances and the barriers of vegetation and minor topographic undulations and constructional features of landscape may not check deflation.

(*v*) Vegetation

In the areas of natural vegetation—largely forests *i.e.*, the Himalaya, the *Bhabar, Tarai,* N. E. Hills region and the scarps and hills of Peninsular India there are two types of land undergoing soil erosion : (*i*) the land actually under forests and (*ii*) cultivated clearings. For the cultivated clearings the conditions and problems are similar to what they are in the cultivated tracts of the N. Indian plain, the plateau or the coastal plains with some minor differences. One of these differences is the comparative protection against wind erosion. The other is higher (other things being equal) moisture content of the soil, because of greater humidity, lesser evaporation, and shade etc. The forested areas themselves are relatively protected against soil erosion in spite of the commonly steep slopes they occupy. Thus the Himalayan slopes under forests or the Western Ghats or the other scarps of Peninsular India are comparatively free from soil erosion. Gullies and ravines are negligible and sheet erosion is also guarded against by the protection afforded by the forests. The forests diminish the force of rain by the foliage of trees, the roots of trees and plants bind the soil, the grass cover provides a protective carpet against erosion, the fallen leaves and other forms of decaying vegetation absorb water and reduce the carrying capacity of running water. If there is no biotic interference, particularly excessive exploitation of forests and grasses for plantation agriculture and shifting cultivation as in Assam hills, the forests can well take care of soil erosion in the area they occupy. If forested lands are converted into arable by the felling of trees, such lands because of steep slopes are amenable to very spectacular and serious (gully type) erosion, *e.g.* Chotanagpur, Assam hills and submontane Punjab etc. It is much less profitable to put such lands to agriculture and aid the causation of several associated problems as the choking of channels, flooding of lower areas and gullying of land, etc., than to let them remain under the natural vegetation cover which has its numerous economic and ecological gains.

With the general increase in population and the augmentation of the habitable areas, the proliferation of towns and villages, the multiplication of roads and railways, the rise of industrial colonies and the extension of mining activities, etc., pressure is increasing on the existing forests. This results in the devegetation and exposure of sloping surfaces where water is subject to such impetuous flow that soil, coarser sediments and larger fragments are bound to be washed down. Such deforested surfaces, *e.g.* the *Cho, Bhabar* and Chotanagpur are the areas of spectacular erosion where not only soil erosion but land erosion too is rampant on a frightening scale.

In 1961-62, of the total reporting[1] area in India about 19% consisted of forests and the percentage of cultivated and cultivable land (net area sown plus fallows plus cultivable waste) was about 58. Fig. 3 shows the distribution in the same map of two very important features—the forests and the arable —that are connected with soil erosion. As will be clear from the map, relatively dense forest lands occur (*i*) in the Himalaya—the most thickly forested zone of India—and the north-eastern hills region, (*ii*) the scarps and hills in a roughly rectangular belt between the Godavari and Ganga Valleys and between Malwa and Orissa Coast, (*iii*) the Western

[1] The area for which statistics were received. The total geographical area in that year was 326.29 m. hectares and the reporting area amounted to 298.81 m. hectares.

and Eastern Ghats in Deccan. These forests protect the surface so long as they are there but these lands (because of their steep slopes) are potentially the most erodable areas of the country. If we are serious in soil conservation, one of the

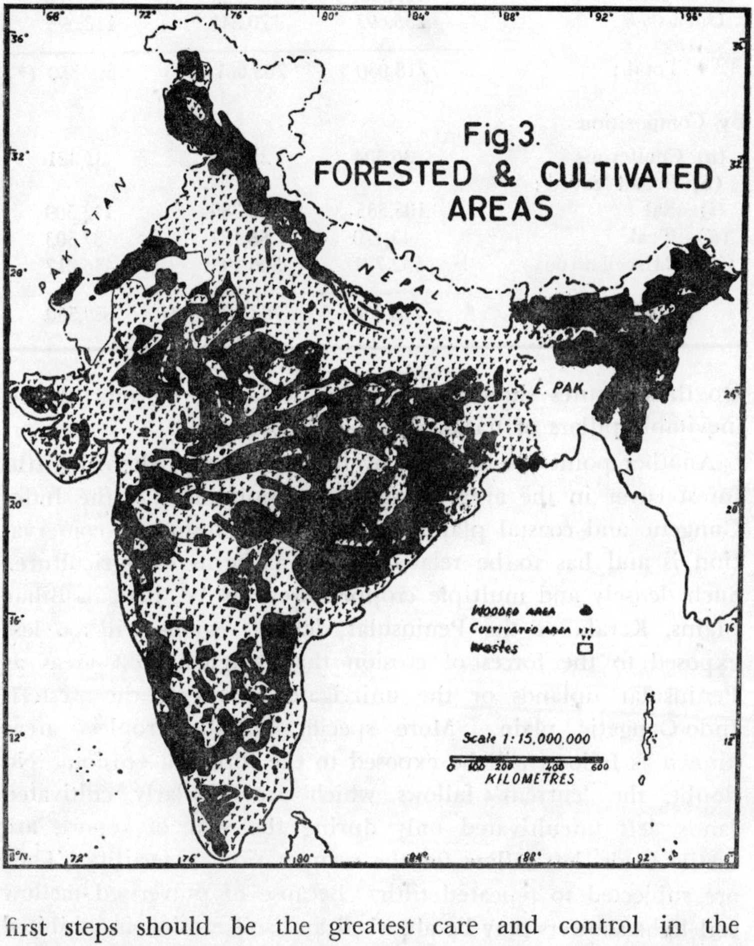

first steps should be the greatest care and control in the exploitation of our best forests. Statistics show that our forests have been contracting and this is indeed serious. The following table makes the point clear :

AREA UNDER FORESTS[1] (sq. km.)

	1950-51	1955-56	1960-61
By legal status			
Reserved	344,405	359,468	316,091
Protected	117,928	168,523	240,572
Unclassed	255,697	170,241	112,095
Total :	718,030	703,661 (**)	689,550 (*)
By Composition			
(*a*) Coniferous	36,304	25,216	32,321
(*b*) Broad-leaved :			
(*i*) Sal	105,535	108,389	113,509
(*ii*) Teak	43,470	58,132	87,503
(*iii*) Miscellaneous	532,721	506,495	456,217
Total :	718,030	703,661 (**)	689,550

Do these figures show that the axe is falling on one of the inevitable pillars of national life and economy?

Another point brought out by the map is that there is little forest cover in the areas of densest population, *i.e.* the Indo-Gangetic and coastal plains and here soil erosion or conservation is and has to be related to the patterns of agriculture. Such densely and multiple cropped areas as W. Bengal, Bihar plains, Kerala or the Peninsular deltaic regions will be less exposed to the forces of erosion than single-harvest areas as Peninsular uplands or the unirrigated parts of the western Indo-Gangetic plain. More specifically, the cropless areas known as fallows will be exposed to the forces of erosion. No doubt, the 'current' fallows which are regularly cultivated lands, left uncultivated only during the year of report, are fertile fields, left fallow for the recuperation of fertility. They are subjected to repeated tilth. Because of pulverised mellow soil light showers may be absorbed and soil erosion is inhibited

[1] 'India, a reference annual 1965' (Delhi, 1965), p. 225.
(*) Includes 20,792 sq. km. for which details are not available.
(**) Includes 5,429 sq. km. for which details are not available.

but such fields are no exception to soil erosion under heavy monsoonal downpours. The other longer-duration fallows, generally occupying sloping terrain or areas abandoned after shifting cultivation are particularly amenable to soil erosion. The distribution of the fallows (*i.e.* cropless lands) is a measure of erosion and is shown in the adjoining map (Fig. 4).

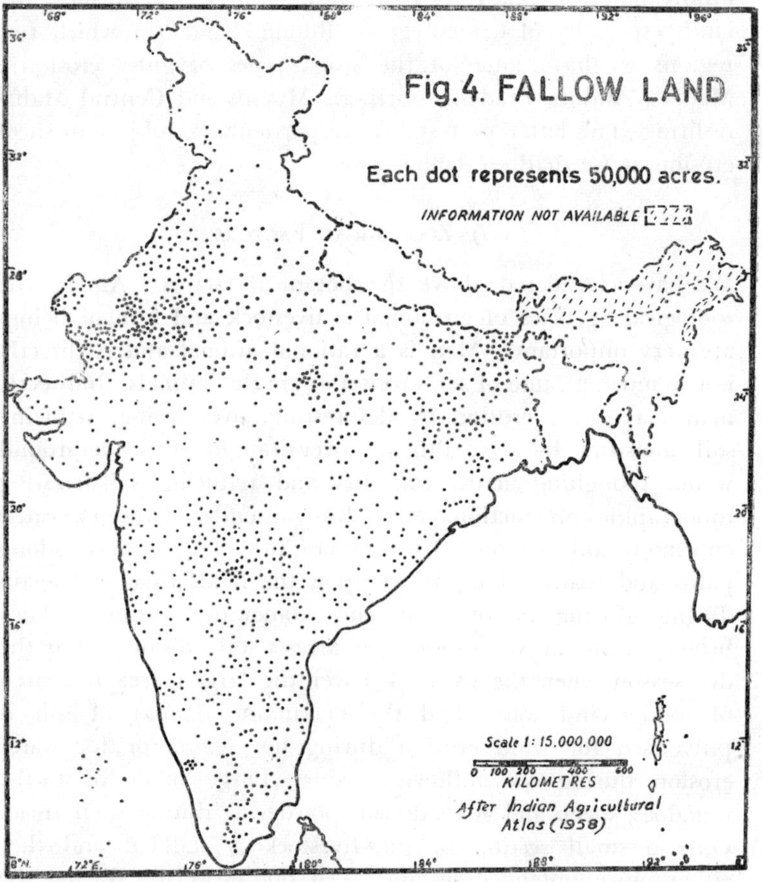

The greatest concentration of fallows in the Indo-Gangetic plain is in the hitherto flood-ravaged Kosi basin. But here a relatively high soil moisture and the rapid growth of grasses inhibit soil erosion. The fallows in the rest of the northern plain are rather evenly distributed. These cropless fields are particularly subject to soil erosion during the torrential downpours of the monsoon. There is a very high proportion of fallows in Rajasthan but the low amount of precipitation is a factor against erosion. Frequently, however, there are short-duration heavy falls causing serious erosion. The proportion of fallows unfortunately is relatively high throughout the Peninsular uplands especially in Chotanagpur, Bundelkhand (in which two regions we have some of the worst types of gully erosion), Madras, Andhra Pradesh, northern Mysore and Central Maharashtra. The latter four states are particularly subject to sheet erosion as we shall see later.

(*vi*) Zoological Factors

We have considered above the botanical factors. Among the zoological agencies of erosion, the livestock and human beings are very important. Man is a rational animal and is directly not so much a cause of soil erosion as are the animals. Indirectly man may cause erosion by deforesting, overgrazing, reducing soil moisture by over-tilth or over-draught of underground water, ploughing across contours and reducing soil-arresting topographic obstructions etc. But untended livestock cause enormous soil erosion. Unlike man they cannot move along paths and roads. They tread upon the fields time and again during grazing or outward and homeward journey. They browse upon the vegetation that arrests soil erosion. It is the dry season when the livestock cover the largest area in search of fodder and water and the maximum amount of soil is pulverised for wind erosion during hot season or for water erosion during the following rains. Large herds of cattle, buffaloes, sheep and goats do soil powdering during their tread. Only a small section of the livestock is stall-fed and they either move untended or are taken out especially during the hot weather when, because of the absence of crops from the fields, the herdsmen have liberty of movement. The density of livestock (Fig. 5), therefore, is a measure of wind and water erosion of soil in India. The map shows that the belts of the highest density of livestock are the Indo-Gangetic Plain, Rajasthan, North Gujarat, the Coastal plains and Madras. In the Peninsular upland there is comparative density in Maharashtra outside the W. Ghats. The distribution of livestock

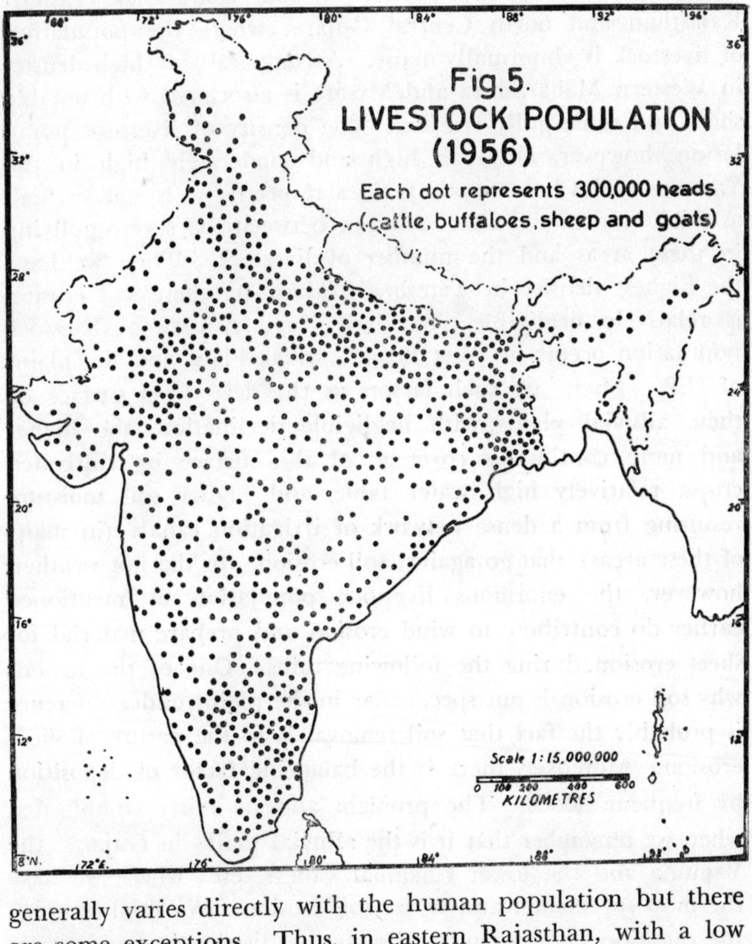

generally varies directly with the human population but there are some exceptions. Thus, in eastern Rajasthan, with a low density of human population, we have one of the highest densities of livestock. Similar is the case with North Central Gujarat. Livestock may cause soil erosion by destroying the vegetation cover and by treading upon soil and pulverising the same. A close view of the map in relation to the areas of marked soil erosion does indicate a good deal of inter-relationship. A high density of livestock in the Western sub-Himalaya co-exists with a severe gully and sheet erosion in the area. Again severe gully and sheet erosion characterise eastern Rajasthan and north Central Gujarat where the population of livestock is abnormally dense. A comparatively high density in Western Maharashtra and Mysore is associated with notable sheet and some gully erosion. The density of livestock population, however, although high and moderately high in the Yamuna and Chotanagpur ravines respectively, is not so high as might indicate a direct relation between the severe gullying in these areas and the number of livestock. Then we have the highest density in Tamilnad (Madras) where soil erosion is relatively negligible. Similarly, high density of livestock population occurs in West Bengal, Bihar Plain and the plains of U.P. There are such factors as the dead level surface of these alluvial plains with negligible declination, the higher and more continuous coverage of the surface by cultivated crops, relatively high water table, and higher soil moisture resulting from a dense network of irrigation canals (in many of these areas) that go against soil erosion. In the hot weather, however, the enormous livestock population as mentioned earlier do contribute to wind erosion and prepare material for sheet erosion during the following rains. One of the reasons why soil erosion is not spectacular in the plains under reference is probably the fact that soil removal is in the nature of sheet erosion. Moreover, there is the balancing factor of deposition by frequent floods. The problem appears fairly complicated when we remember that it is the alluvial plains in Gujarat, the Yamuna and the lower Chambal valleys, etc., where we have the most spectacular and severe soil erosion. We shall examine the question later when we discuss the distribution and causes of soil erosion. It is, however, definite that the density of livestock population is an important factor contributing to soil erosion in India.

Other soil-eroding zoological agencies include mice, ants,

earthworms, jackals, foxes, mongeese, porcupines and a host of insects, etc., which are particularly active during the warm and moist monsoon period when they germinate and work upon the soft soil. It is, however, difficult to indicate their distribution and soil-eroding capacity.

Occasionally human beings do tread upon the fields in search for short distance. This may cause not only the pulverization of the soil but also the channellised flow of water during rains leading to marked soil erosion. The effects of carts carrying manure and harvests and fodder may have worse consequences. Similar may be the effect of working animals moving into and out of fields.

(*vii*) Geological Factor

The country may be regarded as consisting of two types of areas so far as geology is concerned. First, there are those areas where solid geology appears on the surface and, secondly, those tracts, where a layer of soil and sediment occurs between the surface and solid geology. The former type of surface is entirely absent from the Indo-Gengetic plain and largely absent from the coastal plains. They are predominant in the Himalaya and the North-eastern Hills region. They are also dominant in the scarps regions, *e.g.* the Western Ghats, the scarps bounding the different erosion surfaces in the Peninsular upland and the rift Valleys of the Narmada, Tapti and Damodar etc. Bare rocky surfaces are also found on the residual hills, *e.g.* Rajasthan inselberge, Aravalli range, dome-gneisses and other hills of circumdenudation in the Peninsular region.

In the bare rocky regions there may be little soil to erode. As soon as weathered sediment is formed it may be washed down the gravity slopes or higher gradients. Wherever there are gentler slopes and soil accumulation is possible, there is notable erosion if the vegetation cover is absent or poor. In those areas of steep slopes the lithology is immaterial and generally the slope is the overriding factor. Such regions are agriculturally negative or unimportant and the best way of the utilization and conservation of soils in such areas would be to leave them to nature which will cover them with natural vegetation.

The other areas are those where the solid geology is concealed by soil and sediments. Let us examine their broad geological and chemical characteristics in so far as they influence soil erosion. The sticky clayey soils found on the basic Rajmahal lavas or the basic gneisses in Madras or the Deccan lava are less liable to wind erosion than the lighter sandy soils found in the areas of Newer Alluvium in the Indo-Gangetic Plain or the light granitic and gneissic soils of Peninsular India. The presence of salts in the drier parts of Indo-Gangetic Plain helps to bind the soil particles during the hot season against wind erosion. In the notably calcareous belt of north-western Bihar and north-eastern U.P., the large proportion of lime renders the soil friable and aids both wind erosion during the hot season and water erosion during rains. The high content of iron in Chotanagpur and other gneissic and granitic regions of Peninsular India bind the soil as hard mortar against wind erosion during the hot season. The mica schists (by their composition, *i.e.* mica and quartz etc.) yield soils which may be particularly deficient in cohesion and the schist country (*e.g.* Chotanagpur and parts of Mysore), is particularly liable to soil erosion. The shales and slates yield heavier and more cohesive soils than do the sandstones, quartzites, grits, granites and gneisses. Soil erosion, therefore, may be less pronounced in such heavy soil areas as the Chattisgarh plain developed on Cuddapah shales than in the areas of lighter sandy soils, e.g. gneissic and granitic regions of Peninsular India and the Vindhyan and Gondwana sandstones. Moisture lubricates red soils on granites and gneisses as well as those on laterites. Under moisture, therefore, the cementing role of iron is neutralized and the lateritic and other red soils are subject to marked erosion.

The light, coarse and dry submontane alluvial fans coalescing into a piedmont plain, known as the *Bhabar* zone, are more subject to soil erosion than the adjoining moist *Tarai* belt of heavier soils.

The vast sand debris in the Rajasthan desert derived under arid erosion is normally unrelated to the local geology.

(*viii*) Geomorphic Factor

One of the most important factors affecting the nature and extent of soil erosion in India is formed by the topography and surface slopes. The force of water erosion is largely a function of slope which may be, theoretically speaking, anything between 0 and 90°. In the former case there may be little water erosion and in the latter case there will normally be no soil formation. Actually, however, on much lower slope (known as gravity slope) the sediments will normally be pulled down and the slope will be kept bare of soil.

Slopes are so important in soil erosion that it would be indeed incorrect to study soil erosion without trying to understand the implications of slopes and their changes. It is here that geomorphology bears intimately on a human problem. Broadly speaking, India consists of 3 geomorphic units: (*i*) the Himalaya–a Tertiary fold mountain in young to mature stages of the first major erosion cycle, (*ii*) the monotonously flat depositional Indo-Gangetic plain having the characteristics of an epicontinental geosyncline and (*iii*) the polycyclic peninsular upland consisting of several erosion surfaces and inter-connecting scarps.

The Peninsular upland is ever subject to erosion and degradation by geomorphic processes. The result is a regional deloading and imperceptibly slow isostatic uplift. The visible effects of the latest phases of uplift are the raised beaches round the Peninsula. Earlier during the Tertiary and Pleistocene as side-effects of the Himalayan upheavals, the Peninsular upland recorded successive rejuvenations exhibited in the steep scarps that occur as the Western Ghats, the flanks of the Tapti and the Narmada troughs, the Vindhyan Ghats and those surrounding the uplifted peneplains in Chotanagpur and Mysore etc. It is some of these scarps that are the zones of very serious soil erosion in India. The appreciation of the processes of soil erosion and its causes rest on an intimate knowledge of the geomorphic characteristics and processes in these areas. The exogenetic processes, *e.g.* degradation in the upper parts of the slopes and aggradation below are imperceptibly altering the degree of slopes and the rates of soil erosion. The endogenetic processes, *e.g.* dislocation along faults and change of levels involved in warping or tilting etc. might be proved to be perceptible and significant in certain areas.

The second region *i.e.* the Himalaya, is an area largely predominated by steep convergent slopes culminating into snow-capped or bare or plant-covered peaks and ridges on the one hand and ending into narrow valleys (often cultivated) on the other. The major parts of the slopes are forest-covered (Fig. 3). Cultivation is so much limited that the problems of soil erosion are hardly appreciated. Soil erosion is considerable in these regions of India, despite vegetation cover, because of the steep slopes. This can be seen in the enormous sediment load (about 1% of the total annual discharge) brought down by the Himalayan streams. The soil erosion is the result of lofty elevation, stupendous relief and enormous slopes in which respects the region has no parallel in the world. The geomorphic processes tend to keep up the considerable rate of soil erosion. The region is subject to soil erosion partly under glacial and partly under normal conditions. The resulting deloading is probably associated with isostatic uplift attested to on a long-term basis by river terraces and as a short-term phenomenon by frequent earthquakes. Even if the endogenetic processes were too slow to be perceptible, the exogenetic processes are so colossal caused by the compelling degree of slopes that erosion is fairly notable.

The third major geomorphic unit is the Indo-Gangetic Plain. It is a sort of geosyncline and is probably under continued subsidence under the weight of the enormous sediments brought to the area from the Himalayan and Peninsular regions. The endogenetic forces are not clearly understood. Apart from the seismic disturbances causing the changes of levels and affecting soil erosion in the neighbouring areas, probably there are deep-seated upheavals causing such phenomena as the diversion of the Ghaggar and the rise of the Indo-Gangetic divide or the formation of the Yamuna-Chambal ravines etc. The exogenetic causes are continuously active forming levees and other uplands, the choking of river channels, their diversion and distribution, the formation of ox-bow lakes, and other depressions, the formation of 'Bhur' or tracts of sand-hills etc. All these affect soil erosion which, however, is limited by the poor slope of the region.

(*ix*) Sea

The spread of marine water during the normal flow of tides and the ebb of the same may not be associated with any marked soil erosion in the coastal plains and deltaic tracts but storms beat the waves strongly against coastal lowlands when the churning and washing away of soils are inevitable. The mere fact that water is in horizontal motion must cause friction on the land surface and the removal of soil. Now the greater the slope in the coastal areas, the greater the possibility of cliffing and erosion. But the greater the slope, the smaller the extent of the land that can come under the attack of waves and tidal bores or unusually high tides or tsunamis. Usually, however, the coastal regions of India represent the belt where aggradation by streams from land is the maximum and sedimentation or silt-deposition is more pronounced than erosion.

The wave too is a depositional agency on a coast of emergence which is the predominant aspect of the Indian shore. Vast quantities of sand are piled up above the shore line on which the soil-binding casuarina tree is frequently grown. The continental shelf is generally so gently sloping that a slight wave action churns up sediments from the marine floor and piles them along the shore.

There are areas, however, where soil and land erosion by marine action is conspicuous or serious (Fig. 13). There are the stretches of the coast where the monsoon wind blows forcefully at right or nearly right angles to the coast, e.g. the Mysore and Kerala coast on the west and Madras coast in the east.

Marine erosion depends on such factors as (*i*) the lie of the land, (*ii*) the nature and slope of the submarine floor, (*iii*) the fetch, i.e. the length of the sea over which the wind is blowing, (*iv*) the emergent or submergent aspect of the shore and (*v*) the geology and structure of the coastal region. The coast which is under the most serious marine erosion is the Kerala area, particularly Quilon, Aleppey and Calicut districts. Now the question that arises is "Why should the erosion of Kerala coast be so conspicuous?" Broadly speaking, a coast with emergent aspect should be marked by depositional features. The Indian coast can roughly be divided into three categories: (*a*) deltaic coasts, (*b*) the coasts of emergent aspect and (*c*) the coasts of submergent aspect. In the deltaic coasts erosion is ruled out by the dominant rate of deposition that proceeds in the delta mouths. Most of the Indian coast has emergent aspect. This is generally a guarantee against soil and land erosion by waves and currents. In some limited coastal stretches, there are submergent aspects where the coast, *e.g.* Konkan, is rocky and there can be little soil or land erosion within human history. The more vulnerable points may, however, be under vigorous attack.

The S. W. monsoon is a powerful air current lashing marine water against the shore on the western coast. Its force, however, is limited south of the Gulf of Cambay. The Konkan coast, however, is rocky and there is little erosion. On the eastern coast the S. W. monsoon is away from the shore from Cape Comorin to the Krishna delta and further north-east it is parallel to the coast.

The N. E. monsoon is parallel to the coast up to the Krishna delta and further south also it is not as regular a current as the S.W. monsoon. It comes generally in the form of cyclonic storms during November-January. Thus, the most valnerable area is the Malabar coast which has a progradational emergent aspect and the shore region consists of soft alluvium easily liable to wave attack.

CHAPTER III

THE PROCESSES OF SOIL EROSION

In physical conditions the regions mentioned in the first chapter and shown on the map (Fig. 1) differ markedly and lead to marked variations in the manner of soil erosion. Our knowledge of the exact mode of erosion in the different regions is, to say the least, extremely poor. Most accounts in this country are merely qualitative descriptions and no attempt appears to have been made at understanding the exact manner and pattern of soil erosion, not to talk of the quantitative aspects of the problem. The present writer has little technical equipment and knowledge to do justice to such a complicated subject. The following illustrative account will serve only to show how we gloss over highly complicated processes by such terms as sheet, or gully or wind erosion etc.

The Processes of Erosion in the Peninsular Upland

The Peninsular upland generally consists of three types of surfaces: (*i*) scarps and sloping hills, (*ii*) peneplains or plateaus and (*iii*) well-defined valleys. The first category generally has a slope of over 5% while the second and the third categories have average slopes of 2 to 5%. A slope even of 2% is fairly high for imparting such characteristics to the flow of rain water and soil erosion that are very distinct from those in the dead level northern and coastal plain tracts. This sloping or undulating country in the Peninsular upland accounts roughly for about half the total Indian surface. The water flow in such a terrain is what may be called 'cumulative' as against 'distributive' that may characterise a dead level plain (Fig. 6a). By 'cumulative' we mean the growing amount of water during a shower that can be seen passing over the surface as we move from a divide to the adjoining river channel. About 60% of the average village of the Peninsular upland consists of terraced lowland (Fig. 6b) and the remainder as unterraced upland adjoining local hills or divides. During a particular shower erosion in the divide region is limited because of the limited catchment and the small amount of water that flows

FIG. 6(a)

THE NATURE OF RUN-OFF AND EROSION IN AN AVERAGE VILLAGE IN PENINSULAR UPLAND

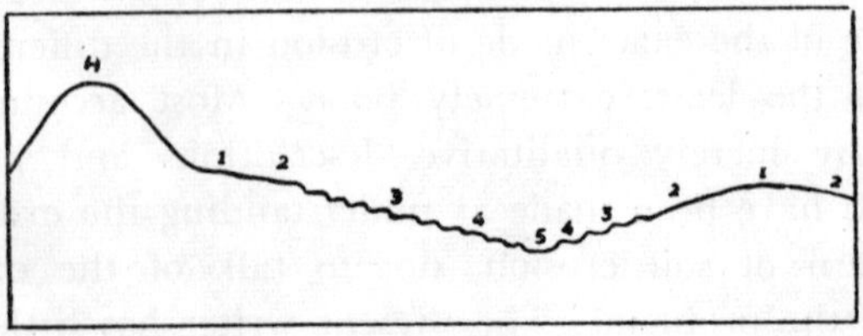

H: Hill, generally soil-less and bare.

1 Rocky slopes and divides, soil-less or pebbly, limited catchment and run-off, soil or sediment removal limited.

2 Discharge and sediment load multiply, erosion on these naturally sloping unterraced fields (sheet or gully type or both) marked.

3 Discharge and sediment load increase but the slope broken by terracing Erosion is concentrated at the off-take points of the terraced fields (see fig 6(b)).

4 The discharge increases to become a sort of sheet flow Erosion is pronounced.

5 Stream

on the surface. Frequently such divide or hill areas may carry some forest or scrub providing additional protection against erosion. But as we move down slope the amount of water increases enormously and soon we notice numerous noisy, impetuous and swift torrents. Such a channelised flow is the direct consequence of the marked slope of the area. In any particular locality, therefore, the situation of a field is more important than the amount of rainfall. From the crest to the adjoining stream there is the inevitable run of water before it merges with the stream flow. In those parts of the Peninsular uplands where the average slope is over 2% one commonly finds during the rains noisy torrents of water at every notable change of level like falls and rapids, and also at the points of

FIG.6(b).

THE NATURE OF EROSION IN THE TERRACED FIELDS.

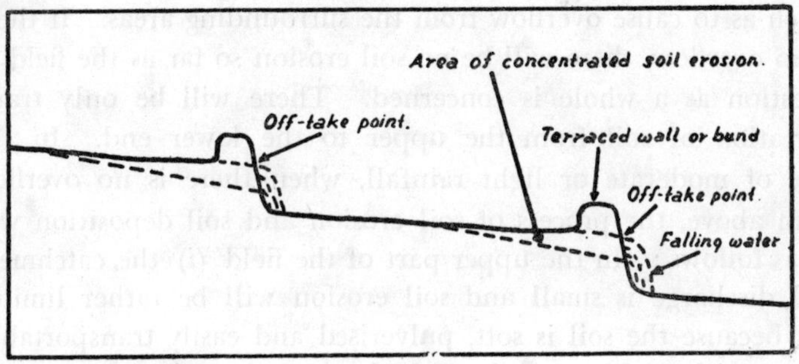

exit in the terraced fields over the lower 'terrace wall' or bund. To certain extent, therefore, the flow of water is in the nature of waterfalls causing gully formation. Thus, the manner of water flow in the Peninsular upland is potentially capable of gully erosion. This will materialize where other physical (*e.g.* faulting, uplift, soil and sub-soil texture and amount of rainfall) and man-made (*e.g.* shifting cultivation and deforestation, overgrazing, etc., and ploughing across contour etc.) conditions aid the incision of land.

In details of mechanism, intensity, environmental setting and other causes, the process of soil erosion will vary from point to point, but there will be some general uniformity in such a process on regional basis. We have examined above some features of the Peninsular region. We may now examine some broad erosion processes in the plains.

Plains

Now let us assume the circumstances when erosion will be the maximum. Such circumstances will include a field bare of vegetation, *i.e.* a current fallow where the soil has been pulverised by tilth ; a heavy torrential rain during the period of S. W. Monsoon particularly when the monsoonal depressions have assumed the nature of violent cyclones. Another contributory factor will be the catchment of the field in question. If the field is separated from the surrounding areas by bunds, the field in question will be the recipient largely of the precipitation falling directly on the field unless precipitation is so much as to cause overflow from the surrounding areas. If there is no overflow, there will be no soil erosion so far as the field in question as a whole is concerned. There will be only transportation of soil from the upper to the lower end. In the case of moderate or light rainfall, when there is no overflow from above, the process of soil erosion and soil deposition will be as follows : In the upper part of the field (*i*) the catchment and discharge is small and soil erosion will be rather limited but because the soil is soft, pulverised and easily transportable, soil particles will move down so long as there is movement of water. The action of beating rain will largely be independent of slope and location in the field and will indeed be a potent eroding factor. Erosion, however, will increase in magnitude further down where the catchment has increased and discharge has multiplied (Fig. 7). In the lower part of the field a

FIG. 7.

THE NATURE OF SOIL EROSION IN THE BUNDED FIELDS OF THE PLAINS WHEN THERE IS NO OVER-FLOW OF WATER ABOVE THE BUNDS.

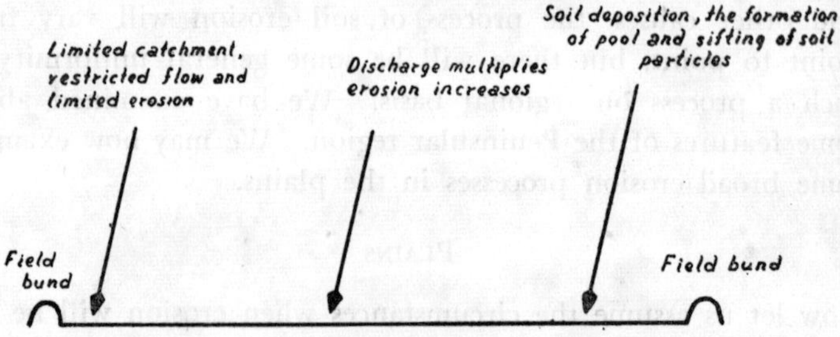

temporary pool, particularly of rain, when rain is falling, may form. This pool will help in the rearrangement or sifting of soil particles. Thus, the total effect of these processes will be erosion of soil in the upper part of the field, removal of the upper soil layer and the appearance at the surface of the lower soil layers and a tendency for the appearance of the coarser soil particles. On the other hand, the lower part of the field will get a layer of fine silt. But for wind, there is no process in nature that will halt and interfere with the process of silt formation in the lower part of the field under the process of water erosion.

Now the other alternative in the case of bunded fields may be overflow of water during heavy downpour or flood. In the areas of very even surface and insignificant slope which characterise the regions of older alluvium, where the bunded fields occur, low to moderate rainfall can be taken care of by bunds and there will be no soil erosion in the sense that all the soils of a field will remain within it. If the surface of the fields were perfectly horizontal, as much rain as is the height of the bunds (*i.e.* 6″ to one foot or so) can be contained within the field without any over-flow and soil transport at all. Some allowance may also be made for soakage, *i.e.* the fields could contain a precipitation of over 6″ to 1′ without overflow.

Actually, however, this is not the case. All areas in alluvial plains have some slope. This will cause cumulative discharge of water so that a field may receive, during heavy rain, more water than it can contain; overflow will result. The overflowing water will carry soil particles from the upper areas, the amount of soil transport being proportionate to the force of downpour, extent of flood and overflow, extent of vegetationlessness of the fields, coarseness of soil, texture and degree of slope etc.

Flood discharge even on plain lands has great force. The momentum of water is the result of the mass of flood water which may be several feet deep. This huge body of water with a weight of several quintals per square metre moving along the direction of slope has great uprooting, digging, grinding and transporting power. One can verify this by actual observation during a flood. The gradient, however, is a very important factor. Its absence will make water stagnant and reduce soil erosion or transport.

After the flood water has subsided to the bund level water will be arrested within individual fields and each plot will get its share of silt that has been brought from higher areas. In the stagnant water the soil particles in solution will get chance to be precipitated as a veneer of silt. This deposition of soil is a process opposite to that of erosion and characterises the phase of subsiding floods. No doubt some deposition particularly of the coarser particles against bunds or other obstructions to waterflow will take place during the rising floods also but silting and soil deposition is particularly marked during the period of subsiding floods.

Summarising, therefore, we might say that the Indo-Gangetic Plains and other plain tracts of India whose fields are bunded (as is generally the case in the clayey paddy regions of older alluvium) soil erosion is insignificant in times of small to moderate rainfall. When, however, there is flooding, which is relatively frequent in these areas, there is overflow of flood water above the bunds and soil erosion will taken place in the period of high flow. During the phase of subsiding floods, however, there will be deposition of silt (as against soil erosion) in such areas.

The plain areas of bunded fields occur in the Assam Valley, West Bengal, Bihar Plain, Eastern U.P. and the Coastal Plain, particularly the delta tracts. These are regions of transplanted paddy and here the annual rainfall is usually above 40″.

The other broad type of areas in the plain will be that of unbunded and open fields. This obviously goes with such crops that do not need the inundation of fields. Such unbunded fields are common in the areas of newer alluvium. Here the alluvium is generally too light and absorptive to be flooded. Such zones occur near the major streams along the levees and adjoining tracts. Here the land generally slopes from the levee highland to the interfluvial depression.

In these newer alluvial uplands or riverine flats the soil is predominantly light being a sandy loam. The unbunded fields are subject to water erosion even when there are light showers. Both the unbunded nature of the fields and slightly higher declination as compared to the older alluvial regions help water erosion of soil on a larger scale than is the case with the bunded fields of the older alluvium. No doubt the light friable nature of the soil does absorb a good deal of initial precipitation during a shower and such absorption will retard surface flow and soil erosion but this is short-lived and soon the pore spaces between soil particles are clogged by air and water and the running water charged with soil particles can be seen in the fields. The erosion is mostly of the sheet type. Water generally flows as a continuous muddy sheet from any micro-regional divide in the directions of slope. Because of the insignificant declination the water flow is 'distributive' and not 'cumulative' as in the Peninsular upland or the Himalayan region. Water flows in several directions from a point. This generally reduces the possibility of large discharge across a surface and guards against marked soil erosion. Yet there is considerable sheet erosion. This can be observed in the field. One may mark after a heavy shower the appearance of the loams of lighter colour in micro-regional upland or divide and the deposition of a dark-coloured silty and clayey layer in the lower ends of the region. This is clear proof of the fact that the most productive top soil is under constant removal from the upland areas. It does temporarily halt towards the lower parts of a region but because of the absence of any obstruction against water flow, *e.g.* embankment, it is later carried away by floods or heavy showers from these lower situations also to still lower levels or to the local streams.

HIMALAYAS

The third broad region as regards erosion is the Himalayan area which is characterised by steep converging slopes. In the same category of surface may be included the steeply inclined surfaces of the Peninsular India, *e.g.* the Western Ghats, and the numerous scarps bordering the plateaus. Here the angle of inclination is generally above 5° and may rise to over 60° or 70°. No soil is normally likely to form at an angle higher than 45° because as soon as the sediment is disintegrated it should be rolled down by gravity, but the fact is that in nature, there may be breaks along surfaces of even higher inclination, although the inclination of the surface, as a whole, may be more than 45°. Now some soil is likely to form in such situation. Another soil-arresting factor is vegetation. Thus, forest trees may arrest soil above them at surfaces that may be much more steeper than gravity slopes. The origin of soil on such steep mountainous slopes is due to the joints and fissures where the sediments washed from sides and above are logged, retain greater moisture, and can grow trees and other vegetation. Indeed most of the stately forest vegetation of pine, deodar or oaks in succession from the south to north as the heights increase that occurs on the Himalayan slopes much steeper than 45°, is due originally to such soil—and moisture-arresting joints and fissures. Subsequently the vegetation itself works as a soil-arresting factor.

The soils on such steeply sloping surfaces, other things being equal, are more liable to erosion than elsewhere in India.

Removal of vegetation plays havoc as it has done in the Punjab Himalayas. Once the vegetation cover is gone there may be large-scale soil removal both as sheet and gully erosion. An idea as to the extent of soil and sediment that is removed from the Himalayas may be had from the data about the Kosi. The total annual amount of sediment getting out of the Kosi catchment, where it leaves the Himalayas (coarse sediment: 16%, medium sediment: 28% and fine sediment: 56%) is about 100,000 acre-feet.[1] This refers only to the sediment in suspension while that rolled down along the bed is probably 15% to 20% of the suspended load. Now the ratio of approxi-

[1] Seshadri, T. N., *Kosi River and its behaviour, Paper* No. 1 (Patna, 1956), p. 10.

mate[1] discharge of water of the Kosi to the total Himalayan drainage represented by the major streams is 8:72. Thus, if the sediment discharge of the Kosi is 1 lakh acre-feet, it is about 9 lakh acre-feet or, roughly speaking, 1 million acre-feet for the whole Himalayan region. This is equivalent to an area of about 1,500 sq. miles (approximately the size of Ludhiana or Jaunpur district) covered with one foot of sediment.

The magnitude of erosion in the steep mountain areas where the average slope is more than the gravity slope can be realized from the fact that in addition to the force of gravity (which is capable of pulling down anything on such surface unless there is obstruction to such motion) there is the momentum of running water charged with the tools of erosion—stones, sediment particles and dislodged vegetation etc. The cumulative volume and force of water at a point on such slope is directly related to the amount of precipitation, the distance of the point from the crest of the divide, the angle of slope and inversely related to the density of vegetation. But for vegetation and cultivated terraces such slopes are always in serious danger of erosion both in the Himalayas and the scarps of Peninsular India. Spectacular soil erosion in such areas where vegetation is absent or thin should, therefore, be rule rather than exception.

In the Himalayas the unbroken length of the mountain slopes may be very large, often several miles. In the Peninsular scarps, *e.g.* W. Ghats or Chotanagpur, it may be several thousand feet. Now the greater the length of the slopes, the greater the amount of cumulative flow across such slopes. The greater the slope, the greater the velocity of water and potential erosion.

"Theoretically, a doubling of the velocity enables water to move particles 64 times larger, allows it to carry 32 times move material in suspension and makes the erosive power in total 4 times greater."[2]

[1] Burrard, Hayden and Heron, *A Sketch of the Geography and Geology of the Himalaya Mountain and Tibet, pt. III, The Glaciers and Rivers of the Himalaya and Tibet* (Delhi, 1934), p. 175.

[2] Buckman and Brady, *The Nature and Properties of Soils* (New Delhi, 1964), p. 229.

The erosive capacity of water, however, is reduced by channellised discharge which characterises all such steep slopes and the relatively dense forest cover occurring in the Himalayas and on the Peninsular scarps.

The Process of Wind Erosion

The land in the vegetationless desert is covered with loose friable and dry soil particles which require only a forceful wind for being deflated. The winds in the Indian desert in January to May are inblowing on its northern and western margins and outblowing to the east, north-east and south-east across the desert. The former blow in from the High that is located over Russian Turkestan. The latter, *i.e.* the outblowing winds are directed to the Low that is located in the south of the Bay of Bengal in January to March, but later, *i.e.* in April and May the outblowing winds of the desert are attracted by the intense heating of the Peninsular uplands in south-eastern M.P. and adjoining areas, which leads to the development of low pressure in the N. E. uplands and Peninsular India and not in Rajasthan as is commonly believed. The Rajasthan low establishes only in June. Thus, in the period January-May, dry winds arrive into the desert from north and west. The immediate area from which they come is the relatively moist irrigated and thickly cultivated Punjab in the north and similar environment in the lower Indus Basin on the west. A part of their south-western component is from the sea which being a maritime wind is relatively dust-free. Thus, the soil particles that might be deposited by the inblowing winds in the desert are much less than the matter that is blown out of the dry and vegetationless desert by the winds going to the east and south-east. Thus, the whole of the Indian desert is subject to wind erosion whose rate is accelerated with the advance of the year. Precipitation from the western disturbances may check erosion but by about March all the contributory factors, *viz.* the speed of wind, amount of temperature (causing desiccation) and drought (associated with the rise of temperature and gradual disappearance of the western disturbances) increase and so much soil is lifted from the ground in the form of dust that everything including land, plants, if any, buildings, men and animals, trains, motors and factories, etc., within the lower part of the atmosphere (a few hundred feet from the ground) is covered with dust and the atmosphere is changed with dust greatly reducing visibility.

Wind erosion is akin to sheet erosion and roughly has a sand-paper action on the ground on which it acts. Within the body of the desert, however, wind has caused the formation of endless series of longitudinal and transverse dunes particularly in the western half to the west of a NE-SW line passing through Churu in the north-east and Sirohi in the south-west. The sand dunes are oriented in obedience to the outblowing winds which sweep across the surface of the desert from south-west to north-east.

While sand is constantly on the move under the influence of the winds which may reach a force of 15 to 25 km. (between 9 a.m. and 4 p.m.) an hour in April and May, the dune area has remained confined to the west of the Aravalli region beyond which it is the finer soil particles alone that have been transported from the desert. But in the north-east there is no physical barrier between the Rajasthan desert and the Punjab plain. Yet the north-eastern margin of the true sandy desert as indicated by historical records has remained where it is for the last two millennia or so. This static position of the north-eastern margin of the desert discounts the theory that the desert is migrating[1] towards the east or north-east. There is however, no doubt as to the fact that large quantities of soil are eroded and removed out of the desert. It is the finer and richer soil in the form of clay particles that is blown out of the desert. The process of wind erosion by hot 'loo' winds covers the vast area that extends from Pakistan border in the west to the Kosi basin in the east, and from the northern scarps of Peninsular uplands in the south to the forested Siwalik and *Bhabar* zone in the north (Fig. 12). Throughout the period from March to June till the onset of S. W. Monsoon, the entire ground in the desert is blasted with the hard summer winds which generally blow from west to east. The period is generally marked

[1] Ahmad, E., *Geomorphic Regions of Peninsular India, Journal of Ranchi University* (Ranchi 1962), pp. 1-29.

by very low relative humidity (below 20% at 5 p.m. in May in the desert), cropless fields, depressed water-table, dried water-bodies, withered and dying grasses, leafless trees and dusty paths and roads etc. All these factors contribute to marked but imperceptible erosion of the richest top soil.

Further removal of these soil particles, which have been removed from their original fields by summer winds, is performed by the early torrential showers of the S. W. Monsoon that transport these soil particles to the river channels and through them to the sea.

The magnitude of soil eroded by the summer winds in N. India can be imagined by the dust coating of all articles in the lower levels of the atmosphere in the entire area from the Indus to the Mahanadi and from the Gulf of Cambay to the Siwalik range. Throughout this area one can see considerable amount of eroded soil lying against all obstructions to the wind, *e.g.* canal, railway and road embankments, field bunds, walls of buildings, mounds, any minor rise in the ground, and trees, plants and stubble in the field. A high wind at a rate of 15 to 25 km an hour ceaselessly blowing from about 9 a.m. to 5 p.m. everyday for about 3 months transports a good deal of top soil. But the process is particularly aggravated at the time of specially forceful winds known as 'dust-raising winds', duststorms and thunderstorms. These occur throughout the area under reference and carry colossal amount of soil particles with them from one area to another. They are erosive in the area of their maximum fury when the wind reaches the gale force of 45 to 75 km. an hour but towards the end of their journey the role of the storms is largely depositional. The path of the storms is generally several hundred kilometers long, although the diameter is usually small, only a few scores of kilometers or less across. These storms cause the removal of top soil from its original position and its final washing off if performed by the early heavy showers of the S. W. Monsoon. The nor'westers, *i.e.*, the summer storms of eastern and north-eastern India are frequently associated with some rain so that some local washing away of the soil particles is also performed by and during the storms themselves.

CHAPTER IV

DISTRIBUTION AND CAUSES OF SOIL EROSION

HAVING EXAMINED the factors and processes of soil erosion in India, it is necessary for us now to examine the geographical distribution of this erosion and the degree of its intensity. The most spectacular and destructive soil erosion in India is of gully type where gullies or ravines of varying magnitude are spreading vertically and horizontally and are eating away land and soil at a frightful rate. The other major form of soil erosion, *i.e.* sheet erosion, may be serious but is relatively unspectacular. Wind, glacial and marine erosion is of regional importance as we shall presently see.

GULLY EROSION

As to the most important form of erosion, *i.e.* gully erosion, we see the following types in the country:

(1) *Very severe gully erosion*

This is the worst type of erosion on Indian surface and occurs in a roughly triangular area formed by lines joining Mathura, Allahabad and Bhilwara, particularly along the Chambal, Jamuna, Betwa, Dhasan, Ken and the Banas (Fig. 9).

(2) *Severe gully erosion*

Somewhat less serious gully erosion than in the area mentioned in (1) occurs in widely separated and ecologically different regions— (*i*) Gujarat in the valleys of the Narmada, Tapti, Mahe and Sabarmati (Fig. 8) and (*ii*) sub-montane section between the Ramganga and the Jhelum in the States of Punjab, Himachal Pradesh and Jammu and (*iii*) areas peripheral to (1). Erosion in these regions is also associated with sheet erosion.

(3) *Moderately severe gully erosion*

Erosion that is less serious than in regions (2) above but is intrinsically quite severe occurs in the following areas:

(*a*) Chotanagpur, (*b*) Mahanadi Valley, (*c*) Upper Son Valley and Mirzapur, particularly Jabbalpur and Narshinghpur disricts, (*d*) Upper Narmada Basin, Upper Tapti Basin, particularly Jalgaon, Buldana, Akola and Amraoti districts and some other tracts in Malwa.

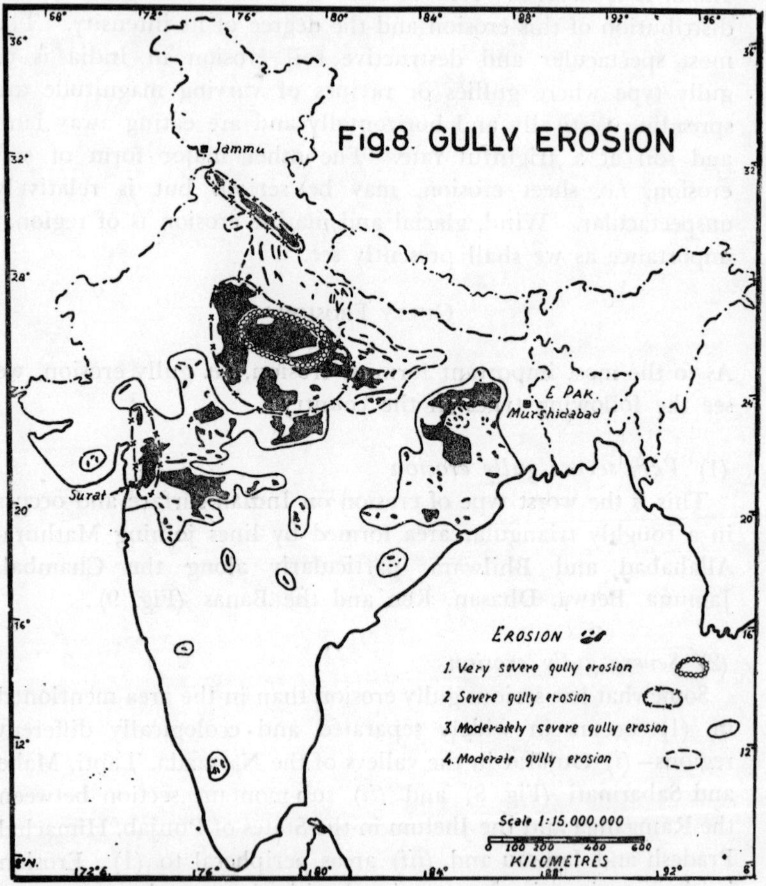

Fig.8. GULLY EROSION

(4) *Moderate gully erosion*

Moderate gully erosion in tracts of the Ganga Plain (Fig. 8), *e.g.* the Valleys of the Hindan and Upper Jamuna, the edges of the Ganga Khadar in western U. P., the valleys of the

Ramganga, Gumati, and Sai etc. and some scattered areas in the Godavari Valley in Maharashtra and Andhra Pradesh and Kathiawar etc.

(5) *Relatively unimportant gully erosion* will be found in most of the areas outside the plains, *i.e.* Himalayas, Peninsular India, particularly on steep slopes.

It is thus apparent from a glance at the map that gully erosion is relatively insignificant in the whole of Deccan, south of the Godavari. It is also insignificant in the plain of the Brahmaputra and that of the Ganga, east of Varanasi. Again, it is relatively unimportant in Punjab, outside the mountain and submontane zone. It is relatively unimportant in Kutch and western Rajasthan.

The distribution of the most notable type of soil erosion (gully erosion) in India, therefore, poses a very difficult problem in physical geography. Why should gully erosion be mostly concentrated in an area roughly enclosed by the triangle whose apieces are Jammu, Murshidabad and Surat ? The intriguing aspect of the question is that the intensity of gully erosion occurs apparently indifferently to geology, elevation and vegetations cover etc. Thus, a good deal of the Chambal-Jamuna zone occurs in purely alluvial country and part of it occurs in the lava region. The Gujarat belt occurs both in alluvium and Deccan basalt. The Punjab zone occurs in tertiary sedimentaries and piedmont gravel plain, the Chotanagpur zone occurs over gneisses and schists etc.

Again, the gullied areas extend from near sea level in Gujarat and Baitarani Valley to about 1000-3000 feet in Chotanagpur and Maikal Range area and to about 3000-4000 feet in the sub-Himalayan zone. The frightening chasms and gullies occur in tracts stripped bare of vegetation, *e.g.* Gujarat, the Chambal and Jamuna Valleys and also in the midst of forest, *e.g.* Chotanagpur, Upper Son Valley and the western sub-Himalayas.

We have, therefore, to look for some cause which is independent of geology and structure, vegetation cover and topography, if we have to give a satisfactory explanation for the force behind remarkable gully erosion in India in a central triangular belt (Surat-Murshidabad-Jammu). It will indeed be wrong to say that geology and structure, topography and vegetation have not affected erosion in India. We have emphasised the importance of these factors. What we want is that we must have some common physical denominator in the different areas of gully erosion. As, for instance, it is easy to say that recent deforestation is the cause of the severe erosion in the Chambal and Jamuna Valley, but it is difficult to substantiate such a thesis. First, there is nothing definite and positive to show that this part of India was wooded in the past, especially when it was so close to the civilized heart-land of India and along the corridor of movement between Northern India and Western Deccan. There is nothing particular in the composition of the alluvium here that should make this zone the frightening wilderness of erosion in the country.

Again, there is nothing locally particular in the composition of the black soil further south that should make it a wilderness of gullies and chasms. If aridity along with torrential rain is taken to be the cause, then such rain showers occur in a very wide area extending from Punjab to Mysore.

Again, conditions in respect of height above the sea, forest cover and geology in the (i) Upper Narmada and Tapti Valleys are akin to what they are in the *(ii)* Upper Godavari and Krishna basins, but gully erosion is much less serious in the latter than in the former areas.

Western Sub-Himalayan Area

Another area of concentrated gully erosion in a localised tract occurs in western sub-Himalayan tract. Spectacular erosion occurs in the Siwalik foothill zone and the adjoining *Bhabar* tract in a belt that is about 50 km. in average width. It occurs west of Ramnagar and extends westwards up to the Jhelum. The annual rainfall is rather uniform about 49 inches near the Jhelum to about 45 inches near the Ramganga. The hilly area till recently was densely covered with pine and other forests and still carries considerable woods. The area consists of closely packed folds of relatively weak rocks (partly uncompacted rocks of the Siwalik belt, the intensely compressed or shattered rocks of the Lesser Himalays further north and the loose shingly and gravelly deposits of the *Bhabar* zone to the south), but this structure occurs east of the Ramganga also. No doubt, the rainfall increases gradually towards the east, but the increase is gradual. Again, the density of vegetation also increases gradually and there is no abrupt dividing line along the Ramganga as regards rainfall and vegetation. There is no abrupt change in the density of rural population and the intensity of landuse. The sub-Himalayan erosion belt runs along a well-defined axis between the Jammuna and Ramnagar (in Naini Tal district) and is coincident with the *Bhabar* and Siwalik and 'Dun' belts.

The most basic cause of marked erosion in this area is probably a notable tectonic upheaval. We know that the Siwaliks represent a former deposition belt which was folded in Pleistocene. The present Indo-Gangetic Plain may meet a similar fate in geological future. There is a tendency for the continuance of uplift in the orogenic belts and of the adjoining depositional areas to be converted into folds. There is geological evidence (*e.g.* the Karewa lake deposits) of the uplift of the Outer Himalayas (Pir Panjal) by about 6,000 feet during Pleistocene. The dismemberment according to some geologists of the 'Indo-Brahm'[1] river appears to be due to a marked uplift in this 'Indo-Gangetic' divide area. It was the continuance of this tendency that probably beheaded the Ghaggar in historic times and diverted its headwaters into the Jamuna. It is probably the continuance of this uplift tendency which in combination with the biotic factors of deforestation and overgrazing and the geological factor of relatively uncompacted and shattered rocks and comparative aridity inhibiting quick regeneration of vegetation that have made this region of ghastly gully and soil erosion that poses serious problems, particularly to the Bhakra dam. Indeed the biotic and geological factors are very powerful here. Uplift is invoked primarily as a factor

[1] Krishnan and Aiyeigar, "Did the Indo-Brahm or Siwalik River Exist"? *Records of the Geological Survey of India, Professional Paper*, No. 6, Vol. 75, 1940.

that induces invigorated down-cutting giving deep gorges and channels into which flowing water from adjoining areas tumbles down and causes rapid and marked land and soil erosion. Uplift also helps in producing steep slopes and powerful water-flow which cause soil erosion.

As to why this erosion is concentrated in the western sub-Himalayas is probably also due to the comparative aridity of the region inhibiting rapid regeneration of felled vegetation. As we have mentioned earlier, the annual rainfall is rather uniform, 40-50 inches from west to east in this belt of severe erosion that extends from near Naini Tal to the Jhelum for about 425 miles in length and about 30 miles in width covering about 12 to 13 thousand sq. miles of area over three parallel Bhabar, Siwalik and Dun belts and the immediately adjoining slopes of the Lesser Himalayas. The intensity of erosion which is here both gully and sheet type increases from east to west and reaches its maximum in the central section in the Valley of the Sutlej, its tributary Soan and the Beas. This is one of the most densely peopled sections of the sub-Himalaya, especially the longitudinal synclinal 'Dun Valleys' (8 to 30 km. wide) that occur behind the outermost Himalayan range, the Siwalik (5 to 20 miles wide). The southernmost zone involved is the piedmont plain formed out of the coalescence of the alluvial fans. This zone is also 5 to 20 miles wide. The alluvial, *i.e. Bhabar* Zone is fully deforested between the Jhelum in the west and the Ganga in the east. The height of the *Bhabar* Zone is 800 feet on the southern end and over 1000 feet at the mountain end. Thus among the different sections of the Indo-Gangetic Plain, the *Bhabar* zone has the maximum gradient. The sediment of this zone is pebble, sand and shingle which make this area extremely porous and friable. The absence of moisture in the soil and sub-soil along with the friability and the lack of vegetation make *Bhabar* particularly erodable.

The Siwaliks occurring to the north have also been seriously deforested. Apart from the damage done to the density and height of the forest growth, the areal extent of the forests has also been seriously reduced leaving only a narrow forest belt

2 to 5 miles wide in the crest zone of the Siwaliks, the lower slopes of the hill both towards the 'duns' in the north and the *Bhabar* in the south having been practically denuded of forests and occupied by settlements, cultivation and other uses of land. Myriads of torrents rush down the slopes from the Lesser Himalaya and the Siwalik crest zone towards the 'duns'. Similarly, such torrents cover the southern slopes of the Siwalik and the *Bhabar*. In their upper sections, the hill torrents (locally called 'chaos') have great slope and deep channels and draw in all the soi from above and their sides in the form of sheets and gullies. In their lower sections both in the 'duns' and the *Bhabar*, these torrents are disproportionately wide channels, choked almost to their brink with sand, shingle and gravels. The enormous quantity of sediments that choke these channels is a measure of erosion that goes on in the upper slopes. The rate of silting in the Bhakra reservoir (Govindsagar) is indicative of erosion in the Sutlej Valley section of the belt. The average annual rate of silting in the lake is 28,000[1] acre-feet or about 4 square miles covered with sediments one foot thick. This has shortened the estimated life of the Bhakra reservoir from 585 years to 390 years.

In the Punjab sub-Himalaya, the cattle population is disproportionately large as compared to the grazing that is available to the livestock of the area. Grazing incidence in acres per animal in the *forested* Punjab, Himachal Pradesh and Jammu and Kashmir is the minimum in India being 1.47, 1.05 and 1.31, respectively[2], whereas the corresponding figure is largest in Assam, Kerala and Orissa ; 142.03, 21.50 and 29.58 respectively.

The vegetation of the area under discussion consists of forests among which the important species is *chir* pine (*Pinus longifolia*). Other important trees include 'toon' (*Cedrela toona*), '*kher*' (*Acacia catechu*), 'shisham' (*Dalbergia sissoo*), 'palas' (*Butea frondosa*), bamboo and 'gular' (*Ficus glomerata*), etc. The common grasses include 'dub' (*Cynodon*

[1] *The Statesman,* March 6, 1966.

[2] Whyte, R. O., *The Grassland and Fodder Resources of India,* ICAR (New Delhi, 1964), pp. 374-375.

dactylon), 'Kus' or 'Kans' (*Saccharum spontaneum*) and *Chrysopogon montanus etc.*

The Chambal Region

As noted earlier, the worst region as regards gully erosion in India occurs along the Lower Chambal Valley and adjoining tracts (Fig. 9). The ravines along the Lower Chambal and the Yamuna below Agra are terrible in magnitude and appearance. One can have the most spectacular view of the erosion in the Chambal Valley by a journey along the Agra–Gwalior route, while crossing the Chambal between Dholpur and Morena or that of the Yamuna near its confluence with the Chambal in Etawah district. The other river valleys that have erosion of almost the same magnitude (only somewhat less) are the Betwa in northern Jhansi, Jalaun and Hamirpur districts, the Dhasan along Jhansi-Hamirpur border, the Sind and the Pahuj between the Chambal and the Betwa. The ravine belts along the rivers are generally 3 to 8 miles wide.

But gully erosion at its worst is to be seen along the Lower Chambal and the Yamuna (Fig. 9) where the gully belt may

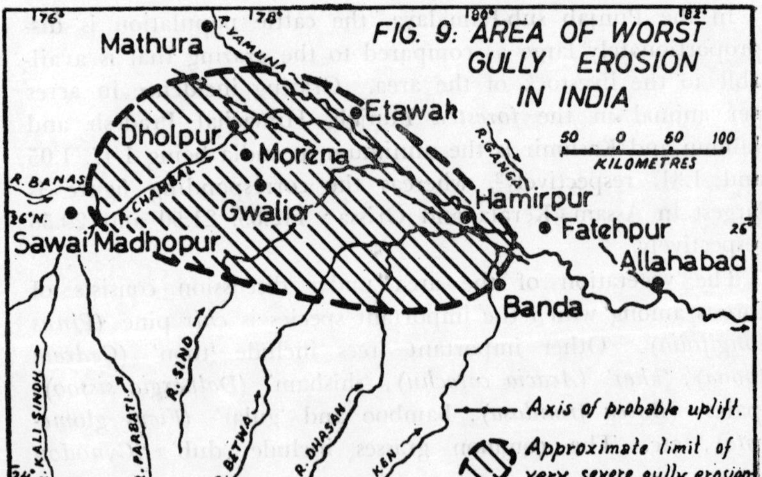

FIG. 9: AREA OF WORST GULLY EROSION IN INDIA

vary from 5 to 12 miles (both sides taken together) along these rivers. The scenery is simply bewildering. It consists of a natural maze where the uneroded columns of land may frequently rise over 100 feet above the level of the stream. There may be grasses clinging to the top or sides of these alluvial columns of land but such vegetation is scanty. In between these alluvial columns are suffocating narrow chasms or gullies, which have provided haunts for the notorious dacoits of the region. Land has been so seriously eaten away by such erosion that there are only tiny remnants of artificially protected pieces of the ground over which are precariously sited the villages of the region. Impassable, dividing, bewildering, and suffocating are proper epithets for the landscape.

The soil of the area is generally clayey. The region except the Yamuna Valley falls in the black soil zone. As we pass towards the south-east, the ravines decrease in magnitude in the rocky and hilly zone. They diminish also as we move towards the east beyond Banda or westwards beyond Swai Madhopur. The worst-affected zone is an oval tract (Fig. 9). The clayey cohesive soil is responsible for the comparative resistance of the narrow high clay columns between the myriads of ravines despite the paucity or absence of vegetation.

It is indeed most enigmatic as to why this part of India should suffer from the worst type of gully erosion. There appears to be some common cause that overrides other factors. There is little appeal in the oft-repeated commonplace argument that erosion is due to excessive deforestation. It is difficult to believe that this area was under a particularly dense forest cover (among the different parts of India) in recent past. This will appear particularly improbable as regards the Yamuna Valley which has been under heavy cultivation and population since the dawn of Indian history. There are many other areas like Bundelkhand, where the percentage of forest area is lower, *e.g.* Bombay or Rajasthan or Punjab, but there is negligible ravining in these areas. The cause may not lie in the character of soil also. Bundelkhand is an area of black clayey soil. The Yamuna Valley is a region largely of loamy older alluvium with a large proportion of saline soils and a narrow ribbon of sandy newer alluvium in the valley flat. Further west towards Rajasthan, the soil is saline and more sandy and partly loessal. Thus, the ravines cannot be ascribed to some common and dominant soil characteristics.

Again, the cause cannot be sought in uniform rainfall conditions. Towards the west the rainfall is around 20 inches while towards the east and south it is about 50 inches. This clearly shows that soil moisture is much lower towards the west, *e.g.* the Lower Chambal Valley than in the east, *e.g.* the Valley of the Ken. This variation of rainfall or soil moisture appears to have little influence on the intensity of erosion which is equally forceful in the oval region defined above.

Again, there is nothing common in the geology of the area that might account for excessive ravining. The northern part of the belt in the Yamuna Valley is readily erodable Pleistocene and recent alluvium. Towards the west, the rock is relatively resistant basaltic Deccan lava. We might indeed say that geology is irrelevant here because ravines do not dig deep into hard rock. There is some force in this statement but it is also true that there are several areas in Bundelkhand and the Chambal Valley where rock does appear in the ravined landscape. By contrast the Yamuna ravines are entirely in the mellow alluvium.

It is difficult to find the cause of erosion in some common agricultural pattern. Fallows are associated with cropless fields where erosion may be especially serious but there does not appear to be any excessive proportion of fallows (Fig. 5) in the region under review, or for that matter, in the other areas of excessive ravining. Statistics do not show any excessive livestock population, *e.g.* cattle, buffaloes, sheep and goats in the region under review (Fig. 5). The treading action of these animals is indeed a powerful eroding force but this does not appear to be particularly excessive in the region under reference.

It is very likely that this excessive and spectacular gully erosion is due to regional uplift which might be extant. The axis of uplift might be along a line passing midway between

Banda and Hamirpur in the east and midway between Hathras and Gwalior in the west. Such an uplift might be ascertained by detailed topographic survey of the areas which might reveal a welt of highland along the line mentioned above.

Continued present-day occurrence of tectonic disturbances in the different parts of India is not an impossibility. Morgan[1] and McIntire have noted uplift and faulting in the Ganga delta in the Madhupur jungle, the surface throw along some faults being more than 50 feet. We shall refer further to the question of uplift but let us examine its influence on ravining and erosion. The most important effect of uplift, so far as erosion is concerned, would be the formation of deep trenches carved by the rivers on the trenched sections of the valley as can be seen in the case of the Chambal and the Yamuna. Such a phenomenon has provided a situation similar to what we see near waterfalls. In its effort to seek its own level rain water tumbles down over the banks which may be as high as 100 feet or more (in the region under discussion) above the level of the stream. The force of such water even where its volume is not large can be imagined. It impinges forcefully on the upper edge of the banks which is a fast-receding knick. Its momentum eats into the alluvium of the bank walls. There is little vegetation that might reduce the force of water or provide protection against destruction. The high inter-ravine ground in the area under reference is partly due to the predominance of the heavy clayey soil, whose cohesive power is relatively high and which prevents rapid crumbling of the inter-ravine space. Coming back to the question of uplift it appears very likely that recent and continued uplift is the chief cause of gully erosion in the area under review. Such a cause has the power of producing results that would be indifferent to all such factors as geology, climate, devegetation and landuse pattern etc.

[1] Morgan, James P. and McIntire, William G., *Quaternary Geology of the Bengal Basin, East Pakistan and India, Bulletin of the Geological Society of America, Coastal Studies Institute,* Vol. 70, pp. 319-342 (1959), Contribution No. 59-2.

It is indeed difficult to indicate the exact cause of such uplift but the symptoms are so definite that one is tempted to consider uplift as the chief cause of excessive gully erosion in certain localized tracts of India. These 3 tracts, *viz.* Gujarat, Chambal-Yamuna zone and Chotanagpur, occur in a more or less arcuate belt (Fig. 10) which is peripheral to Peninsular India and

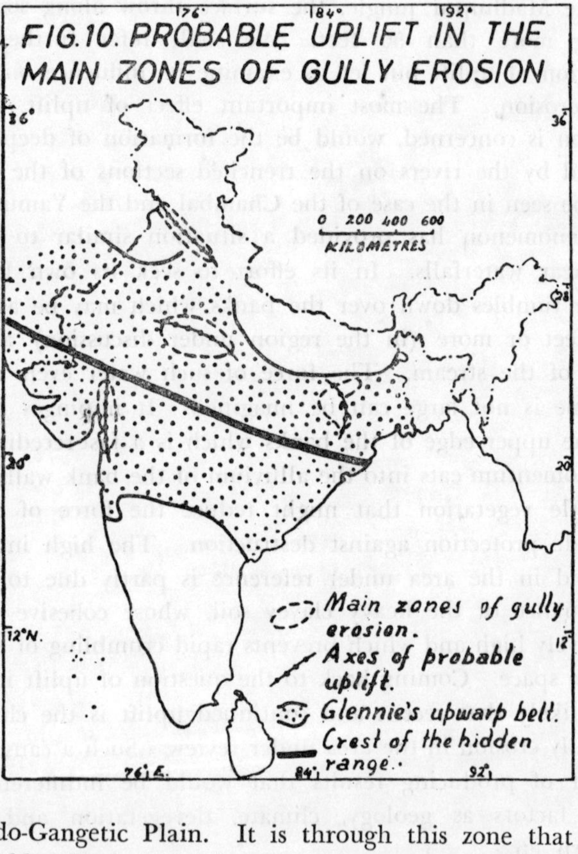

FIG. 10. PROBABLE UPLIFT IN THE MAIN ZONES OF GULLY EROSION

the Indo-Gangetic Plain. It is through this zone that passes the 'Hidden Range' of Glennie, *i.e.*, a zone of greater density[1] and one of an upwarp in the lower layers of the crust.

[1] Daly, R. A., *Strength and Structure of the Earth* (New York, 1940), p. 228.

According to Glennie the zone further north in the Himalaya and sub-Himalaya is a trough. One is tempted to assume that such a sub-crustal feature still continues and the 'upwarp' along the 'Hidden Range' is migrating towards the north raising the area where its axis at present lies. Prof. Muzaffar Ali[1] was inclined to believe that the rigid peninsular block is still pressing against the Himalaya and as a result of this certain areas, particularly where the margin of the shield and the Indo-Gangetic depression occurs are being elevated. He also believed that the uplift is the maximum in the zone where the distance between the Himalaya and the Peninsular upland is the shortest, *i.e.* the area between the Aravalli and the Rajmahals. This theory appears fairly convincing. There is evidence of Pleistocene uplift in Rajmahals, Chotanagpur, Kaimur upland, the Vindhya range and the Tapti and Narmada region, as side-effects of the Himalayan storm. What is needed particularly for excessive gully erosion is the relative rapidity of such uplift. There are several examples of rapid uplift or tectonic dislocation from the different parts of the world. The examples of Madhupur jungle mentioned above, the Allah Bund in Kutch, the formation of a great trench off Moroccan Coast in a few seconds a few years ago are only a few examples.

Ravining, however, is particularly excessive in the central belt in the Chambal-Yamuna region. It is of intermediate magnitude in Gujarat and slightly milder in Chotanagpur and the Ganga Plain (*e.g.* Gumati and the Sai Valleys). Then there are comparative gaps in gully erosion (Fig. 8) in the intervening areas, *e.g.* Bhilwara, Ratlam belt or Ujjain-Dhulia-Ahmadnagar belt or along a line extending from Bilaspur (M. P.) to Daltonganj. The question of peripheral uplift of the Peninsular shield by pressure against the Himalaya is somewhat complicated by this uneven and discontinuous distribution of erosion belts. We are left to the alternative that uplift is particularly marked in the areas of gully erosion—intense erosion—and the rate and amount of uplift is directly proportionate to the intensity of gully erosion. It is difficult to trace

[1] These points were discussed on 18.8.1966.

the cause of uplift. The entire process of uplift is one which is very difficult to measure and its causes are largely in the field of conjecture beyond the reach of scientists. Uplift may be due to thermal changes in the lower part of the crust which would lead to expansion above the areas of excessive heat accumulation (probably by radioactivity) and melting and expansion of the sub-crustal material. Uplift may also be caused by compression caused by the pressure of a foreland or hinterland in orogenic belts as seen in the general Pleistocene uplift of the northern margin of the Peninsular India from Cambay to Rajmahal. Regional uplift may also result from isostatic adjustment caused by excessive regional erosion and deloading. We are neither competent to discuss the cause of uplift in connexion with the subject under study nor will such a discussion be essential. The evidences from gully erosion, however, are so compelling that we are driven to invoke uplift as the main cause of soil and land erosion in the areas under study. We believe in localised marked uplift in (*i*) Yamuna-Chambal. (*ii*) Gujarat, (*iii*) Chotanagpur and (*iv*) Western sub-Himalayan zone. No doubt, in the process of uplift the adjoining areas have also been pulled up and tilted. The probable disposition of the main axes of uplift is indicated on the adjoining map (Fig. 10).

GUJARAT

The erosion belt in Gujarat is an oblong north-south zone extending from the southern bank of the Tapti to about the border of Rajasthan in the Sabarmati Valley. The coastal belt is free from erosion which approximately starts along a line that passes through the estuary-heads of the Tapti and the Narmada. The other river involved is the Mahe in between the Sabarmati and the Narmada.

It is interesting to know as to why this region should stand out as regards erosion. The physical peculiarities of the zone are as follows: The region has a wide range of rainfall from about 50 inches in the Tapti estuary to about 20 inches in the north near Rajasthan border. It occurs mostly beyond the lava belt and falls in the Gujarat alluvial plain. The region has been under close human occupance since early historic times. It has been on the main corridor between North India and West Coast. Consequently, there is no reason to believe that the region has been under forests in recent past and that excessive erosion is due to deforestation. The belt occurs in the midst of an intensively cultivated region. The height of the area above sea level is approximately 0-70 metres.

The basis of our assuming uplift as the cause of the spectacular gully erosion in the area is largely the same as for elsewhere in India. There is no cultural or economic cause which is particularly forceful in this region that has caused the development of hideous ravines. There are other areas of similar devegetation or similar soil or similar cattle population or similar human pressure etc., but there are few areas, none except the Chambal-Yamuna and the Western sub-Himalaya that are so plagued with gully erosion as the area under study. One might say that the estuarine coast in the case of the Mahe, the Narmada and the Tapti should point to submergence of land rather than uplift. The estuarine nature of these rivers is related to the submergence of the past; the present trend being that of uplift. The outermost shore zone between the Mahe and the Tapti is marked by shallow marshy condition pointing to the latest phase of uplift of land.

CHOTANAGPUR

As mentioned earlier, Chotanagpur is another area of spectacular gully erosion. The vertical and horizontal magnitude of gullies or ravines in this region is slightly less than that in Gujarat and western sub-Himalaya and much less than in the central part of the Chambal-Yamuna region. Thus, according to the degree of gully erosion, this region is third, the (*i*) first in order of intensity being Chambal-Yamuna region and (*ii*) the second being the Gujarat zone and the western sub-Himalayan zone.

In Chotanagpur, the distribution of gully erosion has a number of peculiarities. Gullying is more developed on the relatively level or undulating surface of the Ranchi and Hazaribagh peneplains and adjoining belt than on the steeper scarp slopes. Again, if ravining occurs in the wooded zones, it is nearer the margin of the forests than in its interior. The bolder, higher and steeper scarps are generally free from gully erosion because of their general inaccessibility to men and livestock whence the vegetation and forest cover is relatively dense. This is also partly due to the absence of soil and sediment on steeper rocky slopes. It is the gentler and lower slopes of the scarps whose forest cover has been thinned or removed where gullying may be seen.

The worst gullying is seen in the Banka, Jamui and Giridih sub-divisions towards the north of Chotanagpur. Gully erosion of almost equal magnitude occurs further west in Hazaribagh peneplain (Barakar Valley), the steep forested Kodarma scarp overlooking the Ganga Plain being relatively free from gully erosion. The gentler and wider scarps (25-30 km.) that connect the central upper Hazaribagh Plateau (2000′) with the lower northern Hazaribagh Plateau (1000′) are also seriously ravined as the Barakar basin or the Jamui sub-division. Further west, this gully belt continues south-westward into Palamau and Ranchi districts where gullying is equally serious in Chhechari basin, the Upper Sankh basin (in the area called Bhitar Barwe) and Upper N. Koel basin (in the area called Bahar (outer) Barwe). But here too the relatively formidable steep and high forested scarps are practically free from gullying. It is the gentler slopes that are plagued by ravining. All Ranchi peneplain west of the city is plagued by gully erosion in the vicinity of the streams. The environs of Hazaribagh town are gullied like those of Ranchi. To the south-east of Ranchi peneplain gully erosion is conspicuous in Chaibasa plain (Fig. 8) whereas the more densely forested Dalma Range and Saranda and Porahat regions to the south-west are relatively free from gully erosion. Similarly, the densest patches of Palamau forest to the north-west are less ravined. The closeness of erosion is slightly less in Chaibasa than in Hazaribagh and Ranchi.

One can have a close look at the nature of gullying in the environs of Ranchi city in the vicinity of the headstreams of the Subarnarekha (*e.g.* the Kanchi, the Jumar and the Subarnarekha) or in the environs of Hazaribagh town, *e.g.* near the Konar. The headstreams of the S. Koel, N. Koel, Bambhani and Karo, etc., are infested with gullies in Ranchi peneplain.

That gully erosion should be confined to the vicinity of the streams in Chotanagpur, India or elsewhere in the world is because of the fact that the bank or edges of the streams provide the knickpoint on which water tumbles and eats back into the land.

As can be seen from the map (Fig. 8), the eastern part of Ranchi peneplain, the Damodar Valley, particularly its lower section in Dhanbad and the Subarnarekha Valley in Dhalbhum are much less ravined than the parts of Chotanagpur mentioned above.

There is some scattered gully erosion in the Mahanadi basin (Fig. 8) and this may be the side-effects of the basic cause of gully erosion in Chotanagpur. Similarly, in the peripheral areas of the 'cores' mentioned earlier (Gujarat, Yamuna-Chambal area and W. sub-Himalaya) moderate gullying might be related to the basic cause or causes in the 'core' regions.

Now let us examine the circumstances which might account for spectacular erosion in Chotanagpur.

Here, ravines are located on more or less level or undulating peneplains and on gentle slopes, hills and scarps. The steeper and higher slopes because of their inaccessibility to men and animals are densely wooded and are free from gully erosion. Moreover, in these steeper slopes, there is little soil or sediment to be eroded. On these slopes the presence of obdurate rocks is against gully erosion.

Otherwise the gullies occur on the peneplains as well as on gentler hilly slopes. In some cases they penetrate right up to the crest of hills. They occur in the midst of cultivated peneplains and are also found within forested areas, particularly thinly forested zones. There appears to be little relation to geology, because the gullies are equally serious in the schists of Monghyr and Hazaribagh, in the gneisses and granites of

Ranchi peneplain, in the schists and phyllites of Chaibasa plain and in the laterite belt of the 'pat' region. Rainfall varies from about 45" in the north-west to over 60 inches in the south-east and south-west. The relative absence of gullies from the more closely occupied country in Dhanbad suggests that biotic interference is not the main cause. If deforestation were the cause, then Dhanbad would have been most severely gullied. Gullying, as elsewhere, is confined to the vicinity of the streams. The cause appears to be the factors that have deepened the channels of the streams. Such deepened channels would induce rapid and forceful flow of water over the banks and selective gullying would develop in the land behind in those parts where vegetation is poorer, slope greater, soil more pulverised or otherwise erodable.

The cause that can deepen the channels can be uplift that might be extant. Chotanagpur consisting of 3 or 4 erosion surfaces and inter-connecting scarps has been marked by 3 intermittent uplifts from early Tertiary to Pleistocene. This tendency of uplift might be continuing now. The area under uplift is probably the whole of Chotanagpur, the axis of marked uplift being a line as shown in the map (Fig. 10).

Ganga Plain

The Ganga Plain in U. P. is also marked by gully erosion in certain sections (Fig. 8). Such gully erosion is non-existent in the Punjab Plain outside the sub-Himalayan region noted above.

Gully erosion is also non-existent in the Ganga and Brahmaputra plains east of Ghazipur, *i.e.* Gumti's confluence with the Ganga. Gullies are moderate in dimensions and are slightly lower in magnitude than the Chotanagpur ravines. Gully erosion in U. P. occurs along such streams as the Kali Nadi, the Hindan and the Sengar in the Yamuna-Ganga doab and along the Sai and Gumati in the Ganga-Ghaghra doab. Minor gully erosion is also found at the junction of the newer alluvium with older alluvium or the outer margin of the valley flat or *Khadar* of the Yamuna and the Ganga, in western U. P.—between the *Bhabar* zone in the north and Mathura in the south in the case of the Yamuna, and Allahabad in the case of the Ganga. The ravines along the other rivers mentioned above increase in intensity with distance from the source of these rivers. This would suggest that the depth of their channels increases downstream. This is actually the case. In the case of such streams which rise in mountains their deepest streams occur in the hill section, *e.g.* the Ganga, the Yamuna o rthe Kosi etc. In the case of wholly plain streams, the channels are generally shallow near the source, deeper further down and again shallow because of overload of sediments in the lower reaches. This is the nature of such plain streams of the Gangetic Bihar as the Balan, the Burhi Gandak and the Poonpoon, or the Ghaggar in Punjab. But this is not the nature of the streams in U.P. excluding the Ghaghara and its affluents. Here the channels of all streams—the Yamuna, Hindan, Kali Nadi, Sengar, the Ram ganga, the Ganga, the Sai and the Gomati—get deeper downstream. These deep channels appear to be due to the uplift to which we have referred earlier in the case of the Yamuna-Chambal area of worst gully erosion. There it has been suggested that marked uplift covers an oval belt (Fig. 9), the axis of maximum uplift being along a line passing between Banda and Fatepur in the east and Dholpur and Mathura in the west. It is in the northern part of the belt or its fringes that the tract of worst erosion in the Ganga Plain, outside the Yamuna ravines, occurs. Roughly this tract is bounded in the north by an arc passing through Ghazipur, Shahganj and Kanpur. The greater depth of the Yamuna channel as compared to that of the Ganga is probably due to its nearness to the axis of maximum uplift. The channels of the streams close to the Yamuna, *e.g.* the Sengar, are also deeper and the ravines more serious.

Rajasthan

Gully erosion also occurs in the Aravalli region of Rajasthan (Fig. 8) particularly on the western flanks of the Aravalli from near the north of Jaipur towards Punjab in the north-east. This tract falls along the axis (extended westward) of uplift (passing between Banda and Fatehpur in the east and Mathura and Dholpur in the west). There are some other areas of relatively minor gully erosion. These include Andhra Pradesh, particularly the districts of Hyderabad, Nalgonda and Medak, in Maharashtra, such areas occur in the headstream area of the Godavari in Bir and Osmanabad districts on the right and in Nasik, Aurangabad, and Parbhani districts on the left of the river; and further east in the districts of Nander, Chanda and Nagpur in the valleys of the Wardha and Pranthia rivers. To the south-east of Chotanagpur some areas are ravined in Baripada district of Orissa and Midnapur district of W. Bengal. In Mysore there is some gully erosion in the N. W.–S. E. rain-shadow region extending from the district of Dharwar towards Tumkur.

In other parts of the country gully erosion is insignificant. This is particularly so in the northern and coastal plains especially away from the streams. In the Himalayas as well as in the Peninsular region, gully erosion outside the areas mentioned above (Fig. 8) is insignificant. Even the steep slopes of the Himalaya or the scarp zones of Peninsular India are free from gully erosion. The peneplained surfaces of the peninsular tract are also ravine-free. No doubt some of these surfaces are densely wooded; others are deforested and bare of vegetation cover. Some are of high steep slope; others are of gentle inclination. These gully-free tracts are varied as regards lithology, structure, altitude, climate particularly rainfall and temperature; depth, texture and structure of soils, intensity of landuse, density of livestock and human population and intensity of cultural occupance etc. But they are alike in being free from gully erosion driving us to the conclusion that there is some common cause in the ravined tracts and this common cause appears to be uplift.

While gully erosion is not seen in the areas outside the tracts specified (Fig. 8), it can be visualized that academically speaking there is no surface subject to rainfall and waterflow that is completely free from gully erosion. Inequalities of surface are bound to occur in all areas even in the most monotonous plains, not to talk of mountain and plateau areas. In all areas of such topographic inequalities, particularly in the neighbourhood of streams or runnels gully erosion occurs but it is relatively insignificant.

Sheet Erosion

While gully erosion is very spectacular and impresses the eye by its magnitude and hideousness, sheet erosion is also quite serious in India (Fig. 11). Such processes as chemical and

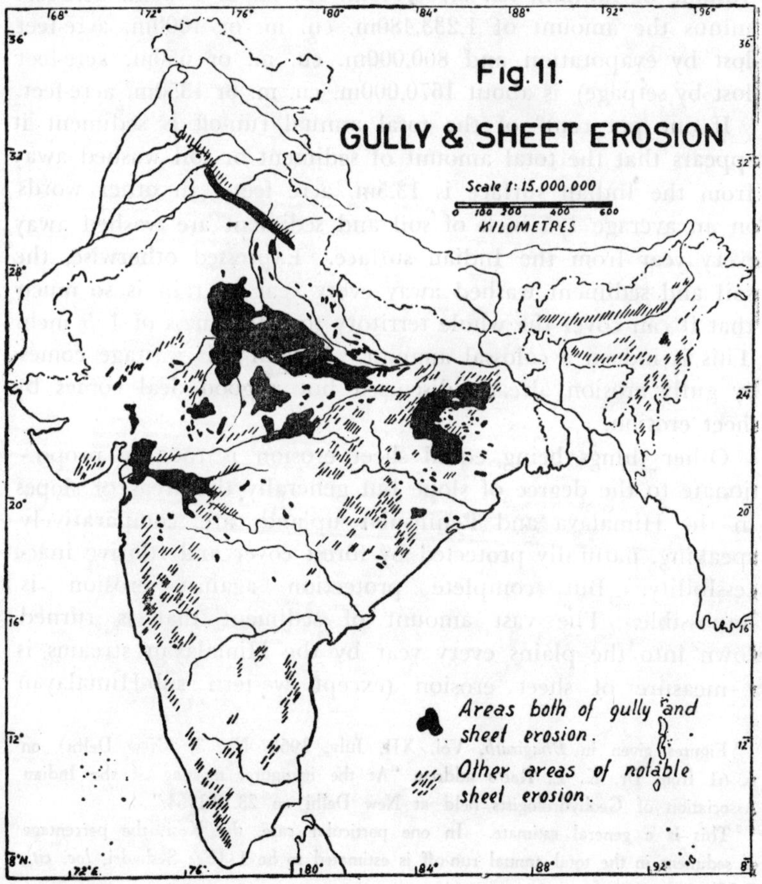

Fig. 11. GULLY & SHEET EROSION

mechanical weathering of rocks into sediments, the removal of forests and vegetation depriving the soil of their binding effect and exposing it to the ravages of rain, the treading of the soil by human beings, the modes of transport, the movement of animals and the work of the plough etc., so to say, prepare the soil for sheet erosion especially where the land is sloping. The magnitude of such erosion can be imagined from the following facts :

It is estimated that the total amount of run-off[1] on Indian surface or the total discharge in rivers of the country (total volume of rainfall, *i.e.* 3,700,440m. cu. m. or 3000m. acre-feet minus the amount of 1,233,480m. cu. m. or 1000m. acre-feet lost by evaporation and 800,000m. cu. m. or 650m. acre-feet lost by setpage) is about 1670,000m. cu. m. or 1350m. acre-feet.

If one per cent[2] of the total annual run-off is sediment it appears that the total amount of sediment or soil washed away from the Indian surface is 13.5m. acre feet. In other words on an average 1/5 inch of soil and sediment are washed away every year from the Indian surface. Expressed otherwise, the soil and sediment washed away every year by rain is so much that it can cover the whole territory to a thickness of 1/5 inch. This is indeed a colossal amount. Part of this wastage comes by gully erosion already discussed but a good deal comes by sheet erosion.

Other things being equal sheet erosion is roughly propo.-tionate to the degree of slope but generally the areas of slopes in the Himalaya and Peninsular uplands are, comparatively speaking, naturally protected by forest cover and rtlative inaccessibility. But complete protection against erosion is impossible. The vast amount of sediment that is turned down into the plains every year by the Himalayan streams is a measure of sheet erosion (except western sub-Himalayan

[1] Figures given in *Bhagirath*, Vol. XII, July, 1965, No. 3, (New Delhi) on p. 61 from Dr. K. L. Rao's address "At the inaugural meeting of the Indian Association of Geohydrologists held at New Delhi on 23.12.1964."

[2] This is a general estimate. In one particular case, the Kosi, the percentage of sediment in the total annual run-off is estimated to be 1 (*See* Seshadri, *loc. cit.*, p. 10).

region where gully erosion is more prominent than sheet wash) that goes on in the Himalaya. Sheet wash is probably most serious in the sloping deforested terrain of Peninsular India but it is present even in the plains. If we stand in any part of the Gangetic plain during a rain shower, it can be seen that the sheet of water that flows across fields is charged with soil particles which are taken down to some position of local base level. No doubt, such sheet erosion may be compensated, taking a long-term view, by *sheet* deposition during floods but until this happens, particularly in the relative uplands where floods are rare or infrequent, sheet erosion, especially in unbunded arable, is bound to affect agriculture adversely.

Sheet erosion is a function both of slope and amount and mode of precipitation. The slope of agricultural land in the bunded arable is bound to affect agriculture adversely.

(1) Indo-Gangetic Plain ·01 to ·2%
(2) Peninsular Peneplains; The margins of the Indo-Gangetic Plain (bordering the Himalaya and the Peninsular Highlands and Coastal Plains) ·2 to 5%
(3) Moderate Hill slopes or Scarp slopes in Peninsular Uplands 5 to 10%
(4) Steep scarp and hill slopes in the Peninsular Uplands; and the Himalaya Over 10%

The annual rainfall varies from over 1000 cm. in the Shillong Plateau or over 400 cm. in the Western Ghats to less than 25 cm. in the Thar Desert and Ladakh. Except in Ladakh and adjoining areas, where it may come as drizzle, the rainfall is generally in torrential showers concentrated mostly during mid-June to September. The sheet erosion capacity of such rain on sloping surface is obvious but it must be clear that on greater steepness, water has a tendency to channelised flow and sheet erosion in such terrain will be local or micro-regional. The sheet wash will first pass on to the adjoining rill or torrent, later to the larger stream and thus from smaller to larger catchments. Such is the situation in the Himalaya and on the steeper hill or scarp slopes of Peninsular region. On the gentler peneplains of the latter region or in the depositional plains, sheet wash acts at once upon relatively large areas.

It may generally be stated that sheet erosion is of the greatest magnitude in the Peninsular region in the unterraced uplands (known as 'tanr' in Chotanagpur) in those tracts where forests have been removed (Fig. 11). It is relatively moderate in the Indo-Gangetic Plain and the Coastal Plains because of slight declination and in the Thar Desert because of poor precipitation. It is important in the areas of gully erosion discussed above because gully and sheet erosion are not mutually exclusive (Fig. 11). Sheet erosion is more important in the Himalayas than in the Plains. The relatively dense forest and vegetation cover in the Himalayas, however, discount the factor of high slope, so that sheet erosion is not as serious as was expected here.

Areas of spectacular sheet erosion in Peninsular India occur in two major belts (Fig. 11). (1) One belt extends along the Western Ghats (both its western and eastern slopes) through such districts as Nasik, Ahmadnagar, Poona, Sholapur, Kolhapur, Belgaum, Dharwar, Nilgiri, Coimbatore and Trichur[1] etc.

The second belt is formed by the 'Peninsular Indian Foreland' extending from the desert near the Aravalli in the west to Rajmahals in the east. Just as the W. Ghats region is one of recent geological uplift and marked slopes where deforestation has been rapid, so this transition belt between the Ganga Plain in the north and the Narmada-Godavari line in the south is one of recent geological uplift consisting of a series of scarps known as 'ghats' where because of the nearness to the densely peopled plains devegetation has been relatively marked.

The third major belt of sheet erosion in India is formed by the N. Eastern Hills including Shillong Plateau, Mikir Hills, Nagaland, Manipur and Tripura. Shifting cultivation resulting in deforestation is the major cause of erosion in the north-eastern hills region. About 25% of the total area of Assam

[1] *Study of Soil Conservation Programme for Agricultural Land*, Programme Evaluation Organization Publication No. 41, (1964, Delhi-6), pp. 73-76 etc.

is reported[1] to be under shifting cultivation. The Forest Department has put a halt to the action of the axe on the reserved and protected forests in Peninsular and extra-Peninsular areas but the forest and scrub that falls outside the total or partial control of the Forest officials and where the villager's axe is free to act, the removal of vegetation has been causing sheet erosion. Generally such lands are located on the crest of the village uplands or on the slopes of the local hills.

Wind Erosion

Wind erosion is quite important in India (Fig. 12). The area suffering from such soil erosion includes Rajasthan desert, most of the Gujarat Plain, north Indian plains, particularly the more friable newer alluvium west of the Kosi and north-western M.P. The arid tracts of Peninsular India in the lee of the Western Ghats, receiving less than 30 inches of rainfall, *e.g.* parts of Maharashtra, Mysore, and Andhra Pradesh also suffer from wind erosion. The problem is particularly serious in northern India (*See* the chapter on the processes of soil erosion).

Glacial Erosion

The area permanently covered with snow in India in the Himalaya is slightly larger than the territory of Kerala. "Out-side the polar regions India has the largest area under snows and glaciers, and the world's largest mountain glaciers are to be found in the Himalayas. The snow-fields, called *himal,* cover about 40,000 sq. km. from Kashmir in the west to Assam in the east."[2] Most of the surface above 5000 mttres (16,405 feet) is covered with ice. The periglacial areas, *i.e.* the areas marginal to the snow-fields are under the destructive action of ice. Every year is marked by a glacial advance during winter (when snow may fall in situations as low as 5000 feet above

[1] *Study of Soil Conservation Programme for Agricultural Land,* loc. cit., p. 197.
[2] *The Gazetteer of India,* Vol. I, Publications Division (New Delhi, 1965), p. 9.

the sea level) and retreat during summer when the ice-covered surface is exposed. The areas actually affected by such oscillation of the ice margin occur between the Karakoram or the Great Himalayan range and the Lesser Himalayan zone. During

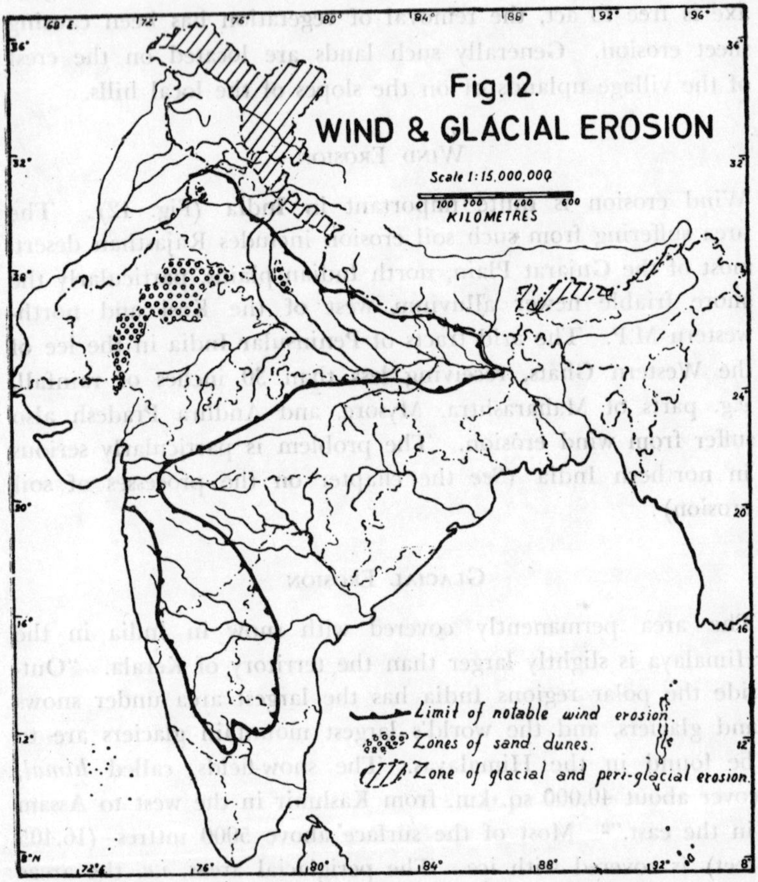

Fig.12.
WIND & GLACIAL EROSION

summer melt the water that issues forth from the glacial margins runs over till or morainic soil that appears exposed. Still lower beyond the winter margin of the ice the soil that is attacked by the rills is the podsolized periglacial soil. It is this latter region that carries some settlements. Some of these are the villages on the alluvial cones at the base of the steep slopes. The adjoining upper slopes are covered with succulent grasses. Such grassy slopes called 'bugial' appear during the summer and are the attraction for the Lesser Himalayans who come here with their sheep and portable habitations in the form of tents.

The heterogeneous soils of these slopes include fragments of different size ranging from boulders to fine silt. The relatively high humus content owing to low temperature and the presence of vegetation is an important merit of the soil. Such erosion as has been described above is, strictly speaking, normal erosion through the agency of water on the sediments and soils prepared by glacial erosion and deposition.

More direct glacial erosion occurs during winter when the snow-fields expand laterally and vertically, and there is a frequency of avalanches which may not only over-run vegetation and settlements but also the terraced fields so that the rather coarse and loose soil is torn away *en masse* from the fields. It is difficult to estimate the extents of such erosion in the Himalaya, but the common occurrence of truncated rocky spurs immediately below the snow-fields is probably a measure of such erosion. The predominant type of erosion performed by the avalanches and glaciers is sheet type (Fig. 11).

Marine Erosion

The waves and currents of the sea have led to the destruction of land and soil from several stretches of the Indian coast. Such erosion of soil is marked by the presence of low cliffs in alluvial coasts. The most notable stretch of marine erosion occurs on the Western Coast from near about Karwar to Cape Comorin with certain breaks. The breaks occur generally where there are off-shore bars between the mainland and the open sea. A second but less marked coastal stretch marked by marine erosion of soil occurs on the eastern coast, south of about 12°N. Here the off-shore bars are generally absent and cliff in alluvium characterises soil erosion. There is a third but tiny stretch of marine erosion near the mouth of the Mahanadi (Fig. 13). Such erosion appears mainly to be

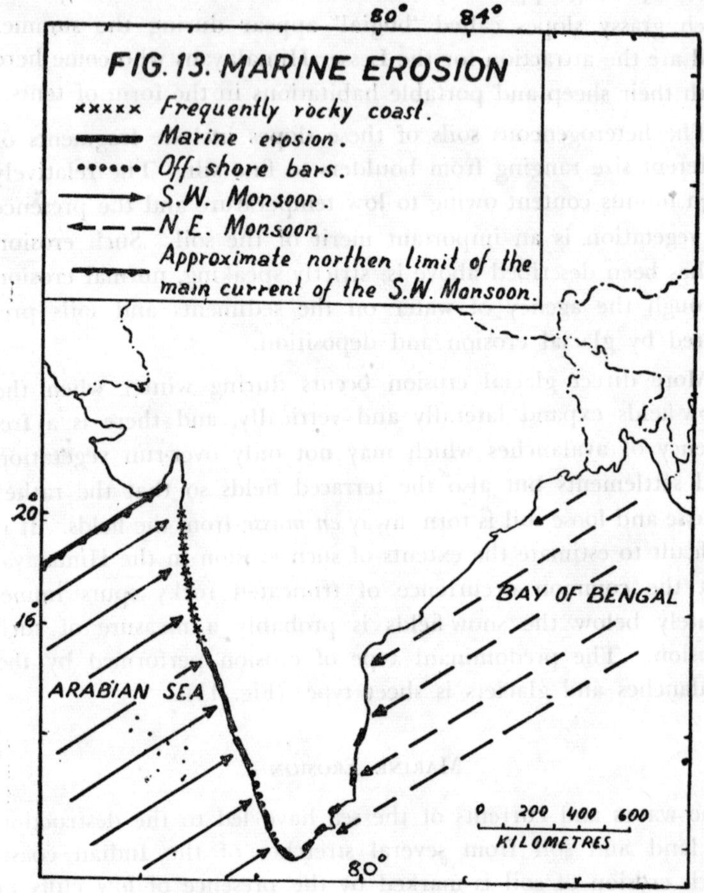

a function of onshore winds. On the Western Coast the force of the S. W. Monsoon is responsible for marine erosion. Further north on the Western Coast there is a predominance of rocky coast so that erosion is inhibited. On the Eastern Coast it is only the southern portion which is under the influence of the N.E. Monsoons and the depressions associated with them. The small stretch near the Mahanadi mouth also has almost ortho-clinal winds. The coast between the Hooghly and the Krishna is parallel both to the S.W. and N.E. Monsoons. The Cambay region is relatively protected by Kathiawar Peninsula and the Monsoon is also a weak current in this area.

The Monsoon causes strong long-shore drifts' causing beach erosion which is aided by tides also.

A graphic account of marine erosion in Kerala appeared in *The Statesman* of July, 13, 1966. The following extracts from this source are very informative:

"Over the past ten years large areas of the Kerala Coast have been swallowed up by sea and during the monsoon months every year, hundreds of square yards of the coast continue to be lost, forcing fishermen to retreat further inland against the advancing ocean." "A few miles south of Alleppey, the highway (between Ernakulum and Trivandrum) is threatened by the advancing ocean which is now only less than 50 yards away at some places (as against over 300 yards in 1957)". "Many of the fishermen's huts which used to dot the coconut groves have vanished into the sea. The Vembanad lake and the Arabian Sea are now separated, at least at one point by only a dangerously narrow strip of land about a mile wide... and...one can forecast the alarming possibility of the sea joining up with the lake in a decade or two....resulting in the irretrievable loss of several square miles in an already land-starved Kerala." "Nearly 200 miles of the State's 350-mile coast line are threatened by sea erosion, the worst affected points being in Quilon, Alleppey and Calicut districts."

Groynes or granite walls have been constructed along only 40 miles out of the 200-mile threatened coast specially at such points "where the sea threatens to swallow the railway lines, national highways, canals, paddy fields, and the ever-narrowing strips of land separating it from the backwater lakes."[1]

[1] *Bhagirath*, Vol. XVII, No. 1, (Delhi, Jan. 1970), pp. 20-24.

CHAPTER V

QUANTITATIVE ESTIMATES

It is essential that we have quantitative estimates about soil erosion. This is necessary for an accurate appraisal of the problem and the phenomenon related to soil erosion. Thus, we require estimates about the area affected by the different types of soil erosion, *e.g.* gully, or wind erosion etc. We also require estimates about the volume of soil eroded. It is difficult to obtain information about both these items even in areas of intensive and quantitative study because of a number of facts. The areal extent of erosion is difficult to find because of the absence of clear-cut margins of such areas and their being intricately mixed with uneroded tracts and the absence of perceptible signs in the land of some forms of erosion, *e.g.* sheet erosion. The difficulties in the estimates of the volume or weight of the soil are too obvious. They are related to the variation in geology, topography, precipitation, vegetation, soil characteristics, biotic and numerous other factors. But the greatest difficulty lies in the fact that it is not known as to where and in what proportion the original run-off and its sediment-load from a region is taken to different destinations inland or into the sea.

Areal Estimate

According to the Planning Commission, 'it has been estimated that about 200 million acres of land, that is, almost a fourth of the country's land surface is suffering from soil erosion'[1]. This appears to be a fair estimate if we cast a glance on the erosion maps (Figs. 8 and 11). This estimate, however, probably does not take into account the area affected by wind erosion (Fig. 12). The figure probably relates to all other types of erosion. The eroded lands fall into different categories. First, there are the gullies, or ravines where both the soil and land has been destroyed and utilization is generally

[1] *Third Five Year Plan* (Delhi, 1961), p. 367.

impossible. Then there are areas of marked sheet erosion, cultivated as well as uncultivated, where the land is intact but soil seriously depleted. Interspersed with the gully-eroded and sheet-eroded land there are patches of cultivation. In the estimate of the Planning Commission given above, it is probably the total area affected by erosion, *i.e.* the land subject to such erosion that it is beyond cultivation and land subject to erosion but sustaining a declining agriculture that are taken into account.

Subsequently,[1] the Commission has provided the following information. The areas badly in need of soil conservation measures are found as *fallows* and include (*i*) the portion of the Peninsular tableland receiving low and erratic rainfall and consisting of the Central Indian Plateau and the Deccan Plateau excluding the forested belt, about 30-40 miles along the W. Ghats and about 100 miles along the Eastern Ghats, (*ii*) the Western Sub-Himalayan and Himalayan region in Jammu and Kashmir, Himachal Pradesh, Punjab and U.P., (*iii*) Chotanagpur Plateau, (*iv*) the areas of red and laterite soils in undulating terrain in W. Bengal and Orissa contiguous with Chotanagpur.

"In the regions noted above, almost all areas except those that are under paddy or receive irrigation need protection. Excluding these (paddy land and irrigated land) areas, the cultivated area in these regions would be about 143 million acres as shown below:

	In million hectares (acres in backets)
1. Peninsular India	
(*a*) Deccan Plateau	37·2 (92·0)
(*b*) Central Indian Plateau	15·1 (37·2)
2. Himalayan and Sub-Himalayan area.	4·2 (10·3)
3. Red and Laterite Soils.	1·5 (3·7)
Total:	58·0 (143·2)."

[1] *Study of Soil Conservation Programme for Agricultural Land* (Delhi-6, 1964), pp. 9-10.

The table refers to the 'cultivated area' falling within the erosion-affected territory. Such arable is mostly fallow or precariously cultivated areas. *The actually negative area, i.e. where not only the soil but land has been destroyed by ravines into highly dissected surface of depressed water table, and which are impossible to cultivate without reclamation and have been abandoned as wastes will obviously be in addition to the 'fallows' or poor arable falling within the ambit of soil erosion.*

Such hopeless negative areas are the ravines or the areas of gully erosion.

Areal Estimate for Gully Erosion

The Planning Commission estimated that "about 3.5 million acres in Uttar Pradesh and about 800,000 acres each in Madhya Pradesh, Rajasthan and Gujarat are badly attacked by ravine erosion."[1] Thus, the ravine-eroded area excluding western Sub-Himalaya, Chotanagpur, Orissa and W. Bengal, according to this estimate, comes to 3·5+2·4 m. acres=5·9 m. acres.

Those who are acquainted with the ravine lands from field observation or from a study of large-scale maps will obviously not agree with the estimates of the Commission, particularly about Madhya Pradesh and Rajasthan where the ravine-affected area is larger.

Let us go into the problem in some detail. According to Bhattacharya[2] the area under ravines in Hazaribagh district of Chotanagpur is 17% of the total. He says: "Erosion has been going on unchecked in this area until last (1957) year when the conservation measures were adopted. A survey shows that during the last 45 years approximately 17% of the total area have been lost to cultivation through gully erosion. This represents the area where there has been complete deteriora-

[1] *Third Five Year Plan* (Delhi, 1961), p. 370.

[2] Bhattacharya, J. P., *Study of Soil Conservation Programme for Agricultural Land* (Delhi, 1964), p. 208.

tion of quality so much so that land use pattern has completely changed. At present the area has fallen into the category of 'uncultivable waste' land full of gullies and some occasional shrub here and there. The remaining 83 per cent of the land has suffered depletion of fertility through sheet erosion."

Our study shows that the extent and magnitude of gullies in Ranchi peneplain is about the same as in Hazaribagh. Gully erosion in south Monghyr and South Bhagalpur is slightly more serious. Gully erosion in Palamau district is less than what it is in Hazaribagh and Ranchi. In Singhbhum it is intermediate between Singhbhum and Ranchi. In Dhanbad it is the least. A rough estimate of the area under gullies in Bihar Plateau or Chotanagpur is as follows:

District.	*% of the total area under gullies.*	*Area in sq. miles.*
Hazaribagh	17	1193
Ranchi	12	916
Singhbhum	10	519
Palamau	7	345
Monghyr	3	119
Bhagalpur	7	153
Dhanbad	3	33
	Total :	3275 (2·096 m. acres)

Our rough estimate of areas under gullies and ravines in the different States are as follows:

Uttar Pradesh (except the Himalayan and Sub-Himalayan districts)

District.	% of the total area under ravines.	Area in sq. miles.
Agra	22	409
Mathura	12	176
Etawah	35	584
Jalaun	24	424
Jhansi	22	854
Hamirpur	24	666
Banda	15	443
Fatehpur	6	98
Kanpur	7	165
Allahabad	7	196
Mirzapur	7	305
Ghazipur	3	26
Jaunpur	3	31
Partabgarh	3	44
Sultanpur	3	51
Azamgarh	2	88
Kheri	2	60
Sitapur	2	44
Lucknow	2	20
Unao	2	36
Badaun	2	40
Meerut	2	46
Bulandshahr	2	38
Moradabad	2	46
	Total :	4890
		(c. 3.129 m. acres).

We have mentioned earlier that the magnitude of ravining is different in different areas (see Chapter IV). Here, we are considering only the approximate horizontal extent of the area under ravines.

In the Sub-Himalayan and Himalayan belt from Jammu to Naini Tal district there is a long narrow zone under ravines and gullies.

Our estimate of area under gullies in the Western Himalayan and Sub-Himalayan zone is as follows:

District.	*% of the area under gullies and ravines.*	*Area in sq. miles.*
Gurdaspur	2	27
Hoshiarpur	10	221
Bilaspur	4	18
Ambala	5	118
Sirmur	3	33
Saharanpur	5	107
Dehra Dun	5	60
Bijnor	4	75
Naini Tal	2	53
Jammu	2	23
Kathua	2	20
	Total :	755
		(c. ·48 m. acres)

In Rajasthan the districts worst affected by gully erosion are those of the Chambal Valley–Dholpur and Sawai Madhopur on the left of the river. Contiguous with these districts are the gullied areas of Jaipur, Alwar and Bharatpur. On the right of the river, the gullied districts are Kotah and Jhalawar. There are some small patches of ravines in the south central districts of Bundi, Banswara, Dungarpur, Chitorgarh and Udaipur (Figs. 8 and 11).

Our estimates of gullied area in Rajasthan are as follows:

District.	*% of the area under gullies and ravines*	*Area in sq. miles,*
Bharatpur	10	313
Sawai Madhopur	10	479
Bundi	2	43
Jaipur	5	270
Alwar	3	97
Kotah	5	240

TABLE—*Contd.*

District	*% of the area under gullies and ravines*	*Area in sq. miles*
Jhalawar	2	48
Banswara	1	19
Dungarpur	1	15
Chitorgarh	1	40
Udaipur	1	68
	Total :	1589 (1.017 m. acres).

In Madhya Pradesh the districts with extensive ravining are the Chambal Valley districts of Morena, Gwalior, Bhind, Shivpuri and Datia. The other affected districts are Sagar, Damoh, Jabalpur and Ujjain (Fig. 8). There are smaller patches of ravines in most other districts of the State. The estimates are as follows:

District.	*% of the area under ravines.*	*Area in sq. miles.*
Morena	15	674
Gwalior	8	160
Bhind	15	258
Shivpuri	5	199
Datia	4	31
Sagar	4	158
Damoh	5	141
Jabalpur	5	196
Ujjain	1	24
Narsinghpur	1	20
Hoshangabad	1	39
Raisen	1	33
Seohore (Bhopal)	1	36
Dewas	1	27
Indcre	1	15
Dhar	1	32
East Nimar	1	41
W. Nimar	1	52
Jhabua	1	26
Ratlam	1	17

TABLE—*Contd.*

District.	*% of the area under ravines.*	*Area in sq. miles.*
Mandasaur	1	40
Shadol (Sidh)	1	54
Rewa	1	25
Satna	1	28
Panna	1	27
Chatarpur	1	34
	Total :	2287 (1.46 m. acres).

In Gujarat the districts worst affected are near the Gulf of Cambay: Surat, Broach, Baroda, Ahmedabad, Kaira, Mehsana and Palanpur. The estimates are as follows:

District.	*% of the area occupied by ravines.*	*Area in sq. miles.*
Broach	10	299
Baroda	10	296
Surat	10	486
Ahmedabad	5	173
Kaira	2	52
Mehsana	2	86
Palanpur (Banas Kanthas)	2	81
Rajkot	2	92
Surendarnagar (Jhalawar)	2	80
	Total :	1645 (1.056 m. acres).

In Maharashtra, the districts worst affected by gully erosion are those of the Tapti Valley: Dhulia, Jalgaon and Buldana. In the Upper Godavari Valley, there is considerable ravining in the districts of Ahmadnagar and Aurangabad. The other affected districts are Bir and Parbhani in the Godavari Valley and Nander, Chanda, Yeotmal, Amraoti and Nagpur in the Valley of the Wardha river. The estimates are as follows:

District.	*% of the district under ravines.*	*Area in sq. miles.*
Dhulia	5	251
Jalgaon	5	229
Buldana	3	113
Ahmadnagar	2	132
Aurangabad	1	63
Amraoti	1	47
Nagpur	1	38
Yeotmal	1	52
Chanda	1	92
Bir	1	43
Parbhani	1	49
	Total :	1109
		(.710 m. acres)

There are considerable areas in Orissa under gully erosion. The estimates are as follows:

District.	*% of the district under gullies.*	*Area in sq. miles.*
Phulbani (Baudh Khondmals)	2	86
Puri	1	40
Cuttack	1	42
Sambalpur	1	68
Dhenkanal	1	42
Baripada (Mayurbhanj)	1	40
Koraput	1	99
Balasore	1	25
	Total:	442
		(.283 m. acres)

In West Bengal, the area under gully erosion is relatively small and the estimates are as follows :

District.	*% of the area under gullies.*	*Area in sq. miles.*
Purulia	3	72
Midnapur	1	53
Bankura	1	27
Burdwan	1	27
Birbhum	1	18
	Total :	197
		(.126 m. acres)

There are some ravines in Coimbatore district of Madras. Elsewhere gully erosion is relatively insignificant. From the above-noted estimates of the areal coverage of ravines, we need not be misled as to the magnitude of gully erosion. By the magnitude of gully erosion we mean (*i*) the vertical relief of ravines and (*ii*) their closeness. The greater these two elements the more serious are the ravines. As such we have already mentioned the ranking of the ravine zones in India. The worst ravines and gully erosion are associated with the Chambal-Yamuna region (Fig. 8). Next in magnitude are the Gujarat and Western Sub-Himalayan zones. The third in order of intensity is the Chotanagpur region. The areal extent of the ravines on the other hand is a different question. This we have roughly estimated on the basis of large-scale maps, field observations and other sources of information. The total extent of ravines or gullied land in India on the basis of the above-noted estimates is as follows :

	Region or State.	*Area under ravines (in sq. miles).*	*In m. acres.*
1.	Western Sub-Himalaya (Jammu & Kashmir, Himachal Pradesh, Punjab and Sub-Himalayan U.P.)	755	.483
2.	U.P.	4890	3.129
3.	Madhya Pradesh	2287	1.464
4.	Rajasthan	1589	1.017
5.	Gujarat	1645	1.056
6.	Maharashtra	1109	.710
7.	Orissa	442	.283
8.	W. Bengal	197	.126
9.	Bihar (Chotanagpur)	3275	2.096
	Total :	15,489	10.364

Thus, if all the gullies of India were put together, they will form a hideous landscape covering some 15 thousand sq. miles, or about ten million acres of land, or an area equal to the territory of Kerala State, or the area which is normally occupied by maize in the country or about one-third that under wheat or about four times the acreage under jute or double that under sugar-cane.

Areal Extent of Sheet Erosion

Gully erosion is more impressively visible in the landscape than sheet erosion. While the former has the nature of comb or furrow action, the latter is illustrated by sand-paper action. While the former is associated both with soil and land erosion, the latter is marked only by soil wastage. It is indicated by the falling output of crops and the decreasing thickness of soil cover. As for example "it is estimated that in Sholapur district of Maharashtra State, 17% of land of medium depth (more than 18 inches) deteriorated into shallower soils (less than 18 inches) in the 75 years from 1870 to 1945. On the Dry Farming Station at Sholapur, it was observed that on deep soil, the grain yield of jowar was 242 lbs. per acre while it was 169 lbs.

in medium soil and 69 lbs. in light soil (less than 9 inches depth)."[1]

No surface which is exposed to water-flow is free from sheet erosion but the area in India which is characterised by marked sheet erosion is about the same as that under gully erosion, *i.e.* about 10 million acres. Such lands are mostly found in the following areas (Fig. 11) :

(*i*) The western part of Peninsular Upland extending from Upper Godavari basin in the north to Tinnevelly district in the south. The deforested back-slopes of the Deccan table land with less than 50 inches of rain form a more or less continuous belt of sheet erosion. The notable districts are Poona, Satara, Sangli, Kolhapur in Maharashtra, Dharwar, Tumkur and Mysore in Mysore; and Nilgiri and Coimbatore in Madras. Further south the areas affected occur on either side of the Cardamom Hills in Palghat, Trichur, Ernakulam, Kottayam and Quilon district in Kerala.

(*ii*) The second conspicuous belt of sheet erosion occurs along the Eastern Ghats from the Palni Hills in Madras through discontinuous patches over Kalrayan, Shevaroy, Velikonda, Nallamalai, Erramala, Bastar and other hills up to the Mahanadi basin near Chattisgarh region.

(*iii*) Areas of marked sheet erosion also occur in Assam hills and adjoining areas in Tripura, Manipur, Nagaland, the main cause being deforestation and shifting cultivation.

(*iv*) The slopes of the Aravallis form another notable area of sheet erosion.

(*v*) But probably the area of worst sheet erosion coincides and overlaps with the region under gully erosion, *viz.* the Chambal-Yamuna area, Gujarat, western Sub-Himalaya and Chotanagpur etc.

Areal Extent of Wind Erosion

High wind period coinciding with rainless season marked by loose arid soil and cropless fields is a feature primarily of

[1] *Study of Soil Conservation Programme for Agricultural Land*, Programme Evaluation Organization, Publication No. 41 (Delhi, 1964), p. 1.

the Indian desert and the Indo-Gangetic Plain up to the Kosi Plain in the east especially in the areas of the newer alluvium. In the areas of older alluvium the clayey soil is cohesive and wind erosion is not perceptible. In the Peninsular Upland, south of the Ganga Plain also the soil in dry weather is generally not loose, friable and erodable. The high content of iron both in the black soil of the Deccan trap and the red and yellow soil of the granitic and gneissic areas render the soil relatively compact during the hot and dry weather, when the winds may be high. Wind erosion, therefore, is a feature primarily of the desert, N.W. Madhya Pradesh, Gujarat Plain and the Indo-Gangetic Plain. Some wind erosion is also experienced in the Deccan belt of loose sandy upland soils in the dry leeward (with respect to the S.W. Monsoon) region in Maharashtra, Mysore and Rayalaseema. Our estimates of the area under wind erosion are as follows:

State.	*% of the State under wind erosion.*	*Area in sq. miles (approximate).*
Rajasthan	75[1]	99,000
Punjab	60[2]	28,200
U. P.	30[3]	34,200
Bihar	15[4]	10,050

[1] The trap belt in the south-east of the State, the rocky and wooded Aravallis, the water-bodies and water-logged areas are relatively free from wind erosion.

[2] In Punjab, the mountain districts of Lahaul and Spiti and Kangra and the surface occupied by water-bodies (canals and their distributaries) or the moist water-logged areas are free from wind erosion.

[3] In western and central U.P. large areas are occupied by water-bodies (canals and their distributaries), and over-moist water-logged areas in canal-irrigated tracts; the rocky and wooded Himalayan belt is free from erosion which is particularly confined to the newer alluvial tracts of the State.

[4] There is little wind erosion in Chotanagpur and insignificant erosion in the clayey south Gangetic Plain. In the N. Gangetic Plain also wind erosion is confined mostly to the newer alluvial area west of the Kosi.

State.	*% of the State under wind erosion.*	*Area in sq. miles (approximate).*
Madhya Pradesh	10[1]	17,122
W. Bengal	5[2]	1,700
Andhra	7[3]	7,420
Mysore	7[4]	5,180
Maharashtra	5[5]	5,950
Gujarat	65[6]	46,800
	Total :	255,6222

This is about 256,000 sq. miles or 20% or approximately one-fifth of the total Indian surface.

Areal Extent of Glacial Erosion

Glacial erosion occurs in the periglacial region especially where avalanches are common (Fig. 11). Although in the Lesser Himalayas snow may occur during winter down to an altitude of 5000 feet above sea level, it can hardly form into huge solid mobile mass that might cause erosion. Generally the seasonal snowfall at the lower levels disappears by gradual melt. In the higher regions where the amount of snow either during seasonal falls or as glaciers is large, avalanches may

[1] Wind erosion is mostly confined to the relatively arid tracts of Madhya Pradesh adjoining Rajasthan, South Western U.P. and Gujarat.

[2] In the districts adjoining Chotanagpur, where the deforested alluvial districts may suffer from moderate wind erosion in the drier upland of light soils.

[3] In the dry Rayalaseema districts where the soils are light, and rocky ground is absent.

[4] In the semi-arid lee of the Western Ghats where the annual rainfall is below 30 or even less than 20 inches.

[5] In the semi-arid lee of the W. Ghats where the annual rainfall is less than 30 and even less than 20 inches.

[6] Wind erosion in Gujarat is probably second only to that in Rajasthan. Forest and rocky patches are confined to Kathiawar and eastern margins only. The vast Rann of Kutch is an arid desert during most of the year. Water-bodies are relatively scarce.

cause the destruction of soils. Our rough estimates are as follows :

State.	% of the area under glacial erosion.	Area in sq. miles (approximate).
Jammu & Kashmir	2	1720
Punjab	.5	235
Himachal Pradesh	2	220
U.P.	.5	570
Sikkim	2	57
Bhutan	2	360
Assam (NEFA)	1	785
		3947

This is roughly equal to the territory of Tripura.

Marine Erosion

We have discussed in the preceding chapter the causes and distribution of marine erosion. The approximate estimate about the coastal stretch subject to marine erosion is as follows :

State.	Length of the coast subject to erosion.
Mysore	90 km.
Kerala	200 km.
Orissa	100 km.
Madras	20 km.
Total :	410 km.

Vertical Extent of Soil Eroded

We have mentioned earlier that about 1/5 of an inch of soil is washed away from the Indian surface every year. But this is an over-estimate. It is based simply on the quantity of rainfall run-off (after deduction is made for evaporation and soakage) assuming that the sediment load in the run-off is about 1 per cent. Over-estimate results from a number of facts. (*i*) First, the run-off from a particular point or region takes into account the precipitation, evaporation and soakage at that point. As this run-off passes down it may come across dry land under sunny sky and may dwindle or disappear before it reaches local base-level or the sea. Indeed, if precipitation at all places occurred at the same time, the run-off would be cumulative and instead of dwindling it would phenomenally augment itself causing flood. But this is not always the case. It is an exception and not the rule. (*ii*) Thus, the eroded soil that starts from one place decreases (in quantity, and the size of the particles) with the decrease of run-off and may settle down completely with the vanishing of the run-off except in rivers and other channels. At the time of floods the eroded soils spread as silt and settle with the subsidence of the flood. Thus, it is clear that all the soil that appears as suspended sediments in the average total Indian run-off does not go out of the country into the sea.

It would, therefore, appear that our earlier figure of .2 inch of soil and sediment being annually washed away from the Indian surface is an over-estimate. If that were true the total washing away of soil from India in 1000 years would be $1000 \times .2'' = 200''$ = about 16.6 feet or about 33 feet during the Christian era. This is obviously too much and the real picture is what we have given in the preceding para. No doubt isostatic uplift is there to compensate for such wastage and lowering of land; otherwise the sea would have covered many marginal areas of the country during historic times itself, even if degradation by soil erosion were much less than 33 feet in about 2000 years.

CHAPTER VI

SOIL CONSERVATION

THIS BOOK has attempted to present the various aspects of soil erosion in India. Little exact knowledge exists as to the a real extent and quantitative magnitude of the different types of soil erosion. Few have bothered themselves about the causes of erosion. No doubt there is the oft-repeated language on felling, over-grazing and ploughing across the contour, etc., without any attempt at a serious thinking on the relative importance of these factors. Still less is known about the exact processes of erosion. Quantitative work is practically nil. We have, therefore, concentrated mainly on the different aspects of erosion. But it will be relevant to add a chapter on soil conservation in India because one naturally expects that when a problem is highlighted its solutions should also be suggested.

The soil conservation measures are probably better known than the causes and characteristics of erosion. The most spectacular example of soil conservation measure that has been followed in India since time immemorial is terracing (generally bench terrace type) which will be found in almost all the villages of Peninsular upland from the Nilgiris to Rajmahals in the north-east and to the Aravallis in the north. Terracing is also seen in most villages of the Himalaya, particularly in more thickly peopled part of the region, *i.e.* the Lesser Himalayan zone north of the Siwaliks. This is a voluntary measure that was adopted by the farmer on realising the immensity of erosion on marked slopes. It must have cost labour that would amount to about the total outlay of one Indian Five Year Plan. It is thus obvious that our farmers are not ignorant of the disastrous consequences of soil erosion.

That terracing of sloping land has been in such difficult terrains as in the Himalaya is a testimony to the farmer's consciousness of the inevitability of conservation. The areas that are now plagued by soil erosion are mostly unterraced lands. This appears to suggest that such lands were in historical past healthy un-ravined cultivated areas, or under forests. Be it as it may, the present state of affairs in the areas of marked erosion call for conservation measures. In Chotanagpur, where Hazaribagh district has been selected for soil conservation programme, the conservation measures adopted are as follows :

(*i*) Construction of embankments parallel to the contours. These barriers against drainage help impounding seasonal run-off into a reservoir of water, and arrest the sediments washed down by sheet erosion from above. Where the embankment runs across the gullies, the latter are automatically plugged by the embankment.

(*ii*) Small masonry and earthen dams across individual gullies have also been constructed.

The adjoining map (Fig. 14) shows the distribution of the 22 districts where soil conservation schemes had been started up to the Third Five Year Plan. It will appear that many of the worst-affected districts, *e.g.* Hoshiarpur, Gwalior, Jaipur, Hazaribagh and Baroda are under soil conservation schemes. Emphasis has also been given to such problem districts as Bilaspur, Koraput and Hazaribagh, etc., which have the catchments of important river Valley projects. Yet the work is largely in the nature of demonstration schemes in most districts and we are only on the fringe of the problem.

The soil conservation measures are well-known. It is the enormous cost which stands in the way of implementing conservation measures. Sometimes the cost may be too much for the return it brings. The only practical solution to the conservation problem appears to be cooperative action by the villagers themselves. Such cooperative action is and has been there in other fields of activities, *e.g.* the construction of village wells, tanks, and roads etc. The Government may grant some subsidy.

In the areas of gullying (Fig. 8) (*i*) ploughing up and cultivating where the ravines are incipient, (*ii*) straw-plugging or earth-plugging where the gullies are minor to moderate and (*iii*) dam-plugging where the ravines are serious are some of the necessary measures.

Generally the areas subject to ravine erosion occur at lower levels near streams and are potentially richer lands but more difficult to save, while the areas subject to sheet and rill erosion

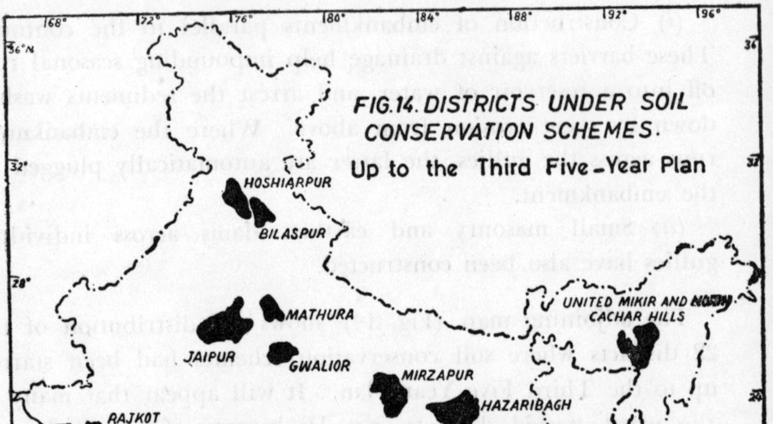

FIG. 14. DISTRICTS UNDER SOIL CONSERVATION SCHEMES. Up to the Third Five-Year Plan

occur on sloping interfluvial uplands which are drier, and have thinner, coarser and less productive soils. In the case of such lands conservation measures are not so difficult as in the case of ravines.

Conservation Measures in the Areas of Sheet Erosion

In the areas of marked sheet erosion (Fig. 11), *e.g.* the 'tanr' or village uplands in Chotanagpur, Maharashtra, Mysore, Kerala, eastern M. P. and in the Himalaya the following well-known conservation measures are of help:

(*i*) *Contour ploughing*: It is interesting to mention here that, so far as the terraced lands in these areas are concerned, it is rare that they are not aligned with contour lines. The lower slopes of the village lands in the local river valley are usually terraced. It is the upper land 'tanr' which is relatively level and where the slope is not so marked as to necessitate terracing. Here unconscious of the direction of slope which is not marked, the farmer may sometimes plough the land across the contour, *i.e.* along the slope. The furrowed channels will then greatly aid the flow of soil with running water during rains.

(*ii*) *Strip cultivation*: It is another conservation method in such areas. Here the slope may be covered by relatively narrow strips of ploughed cultivation, *e.g.* maize or potato with those of other crops. The ploughed strips will check erosion as will the non-ploughed strips growing such grains, as autumn rice or *marua* and other millets or pulses.

(*iii*) *Avoidance of fallowing and cultivation of lands*: Fallowing involves cropless fields. Efforts should be made to cover the fallows as much with crops as possible. One may ridicule such suggestion by enumerating such difficulties as the absence of water supply in the uplands and the very marginal fertility of such lands which inhibit cropping and make fallowing inevitable. One of the greatest maladies of our society is the high proportion of idle labour. This does not require much proof. In 1961, the total population[1] of India was about 440 m. The number of working population then was (43%), about 189 millions. The number of the working population (both males and females) devoted to cultivation including cultivators and agricultural labourers was c. 70% of the work-

[1] *Census of India*, Paper No. of 1962, 1961 Census, Final Population Totals (Delhi, 1962), pp. 396-397.

ing population, *i.e.* about 132 millions. The total cropped area in that year was about 385[1] m. acres. This gives a ratio of about 2.9 cultivated acres per cultivating worker. The fallows may, therefore, be converted into sown. Now that fertilizers are more easily available and the means of agricultural water supply are increasing, it is possible to reduce fallows and increase cropland which will be one of the chief means of checking sheet erosion in the uplands.

(*iv*) The construction of *waterways* between the strips noted above will greatly help soil conservation in the areas under discussion. The soil that may be sheet-washed from above may partly be arrested in the drain so formed. It may increase soakage of the soil. Such ditches should preferably be covered with grasses less they initiate gullying.

(*v*) *Contour bunding*: If the slopes are marked as is the case in most of the areas under reference, contour bunding has to be resorted to. In most of the Peninsular districts where sheet erosion is prevalent, contour bunding has been adopted in the districts subject to soil conservation measures (Fig. 14). This is a sort of terracing where only bunds are thrown up at suitable intervals parallel to the contour and there is no need of forming a level terrace between the two bunds.

(*vi*) If the methods noted above are not effective, *terracing*, as has been done in the lowlands *don* should be resorted to. Terracing will break the slope into fields of gently sloping land and will help soakage. As the uplands, except in the Himalaya and Peninsular areas of steep slope, are generally of small slope, the terraced fields would be relatively large and easier to construct because of the low terrace walls needed. Agriculture on such low-walled terrace will be easier than on the terraces, or on steeper slopes occurring below.

(*vii*) *Horticulture* or bamboo cultivation may be adopted as other soil conservation measures in the Peninsular upland. Such fruit trees as guava, papaya, banana and mulberry grow rapidly and easily flourish on upland soils if they are protected by some irrigation and against animals. They can form into

[1] *India Pocket Book of Economic Information* (New Delhi, 1966), p. 33.

orchards within one to three years and be more paying than traditional upland agricultural yields. Likewise, bamboo clumps may pay more than traditional agriculture and along with their ample leaf mould will protect soil against erosion.

Conservation Measures in the Areas of Gully Erosion

(*i*) *Ploughing up of incipient ravines*: On the margins of ravine zones throughout the country the gullies are in incipient stage. Such areas can be ploughed up and converted into rolling surface. Cultivation and sodding should follow. Ploughing up can be taken only when sufficient or long-continued rain has rendered the soil moist and soft. Such a situation is available during and after S. W. Monsoon or after spell of winter rain. Post-monsoon ploughing may be followed by the sowing of some *rabi* crop, *e.g.* wheat, barley, or some pulses. Early monsoon ploughing may be followed by some autumn crop.

(*ii*) *Straw-dams*: If the ravines are of larger size and cannot be checked by ploughing, dams or plugs of straw should be used. There are a number of rapidly growing hedgeplants and other bushy plants that can be grown across the gullies to act as check-lines below the straw dams. Such plants are: *Putus**, *thethar*+ and bagrer (Jatropha cuccas)**. All these are evergreen quickly-growing bushes and are generally avoided by grazing animals. In all the gullied areas some grasses can be found growing. The upper parts of such grasses can be chopped off for use as straw dams.

(*iii*) When we penetrate deep into the ravine zones the size of the gullies increases. *Earthern* or *masonry* dams or brush dams become inevitable in such cases. The sediment that

* An evergreen thorny profusely branching bushy plant growing wild over most of the Indian Peninsular upland, to an average height of 6-10 feet, frequently used as hedge plant and bearing beautiful pink, whitish pink and deep orange flowers and pea-sized sweet black berries eaten by children.

+ A smooth barked quickly growing, bushy plant with blue bellshaped flowers.

** A quickly growing evergreen smooth-skinned bushy plant with white juicy secretion, used as tooth stick.

comes from above will gradually fill up the gully. Such dams are under construction in the districts where gully erosion occurs and soil conservation measures are under progress.

The conservation measures will vary with the size of the ravines. Gullies have been classified by Bennett[1] as follows:

Small gully—less than 3 feet deep;
Medium gully—3 to 15 feet deep;
Large gully—over 15 feet deep.

According to this classification, the major proportion of the Indian ravines is large. Most of the ravines in the Chambal-Yamuna region, Gujarat and Western Sub-Himalaya are large. The ravines in the peripheral tracts of the zones, the ravines of Chotanagpur and the Ganga Plain in U. P. are mostly medium to large. Small gullies are widely distributed in the other areas of gully erosion (Fig. 8).

Bennett has also given a general classification of the 'contributing drainage area'[2] as follows:

Small drainage area—Less than 5 acres;
Medium-sized drainage area—5.50 acres;
Large drainage area—Over 50 acres.

The run-off through a gully will depend on the contributing catchment. In such closely gullied tracts as the Chambal-Yamuna region, the catchment, except the marginal areas, has been thoroughly dissected and eaten up. It is in the areas of medium and small gullies, *e.g.* Chotanagpur where the contributing drainage area may be relatively large.

Now the conservation measures that may profitably be applied in the different parts of India will be different.

(*iv*) *Fencing off for natural re-vegetation*: In such hopelessly ravined tracts as the Chambal-Yamuna and Gujarat or

[1] Bennett, Hugh Hammond, *Soil Conservation*, Mc-Graw-Hill Book Company (New York, 1939), p. 513.

[2] *Ibid*, p. 513.

western Sub-Himalayan region, one of the important measures that will be helpful is to fence off the areas so that protection against livestock leads to natural re-vegetation. It is difficult to establish vegetation without such protection. Iron posts and barbed wire may provide effective protection. If the cost of such wire, posts and labour is about Rs. 5,000 per mile, a sum of Rs. 200,000 will be needed to fence off an area of 100 square miles. Viewed in terms of ultimate gains, the cost is not much.

(*v*) *Breaking and levelling up of the Ravines*: In India where quick results are essential in any development plans and where there should be concrete incentive for the farmer who reclaims such hopeless lands, it would be profitable to break the inter-ravine divides and fill the gullies with the earth so obtained, carry on contour ploughing and the cultivation of some perennial crop suitable to the region. The ravines in the core areas are too formidable to admit of such a measure but we might attack them from the outer margins where the gullies may be small and medium. Such newly reclaimed and 'loose' land will be subject to ravining particularly on the lower margins where they merge with the unreclaimed ravine belt. Here protection may be sought by marginal bunding and vegetation. Such work can be done only by the villagers and in tiny fragments because of the difficulties involved, but those fragments should be continuous. Isolated reclaimed patches can more easily be ruined by erosion than large continuous patches. Any one who has seen the hideous ravine in the Indian scene, *e.g.* along the Chambal or Yamuna, is tempted to think that these deep chasms, sometimes more than 100 feet, with little inter-ravine land intact, extremely inaccessible for men and animals, are beyond the claims and rights of any one. Even if a settlement survey for checking the rights of ownerships were taken up in such tracts, it would be extremely difficult for the officials to enter all the points. In the present circumstances the ravines either vest with the Government as unclaimed waste called *ghair mazrua* (Arabic, meaning uncultivated) or belong to the villagers for whom it may be difficult to establish their claim of right. Even if a record of right exists, because of the utter poverty and hopelessness of the gullied lands, the owners may be willing to part with them at nominal cost. The Government might obtain such land and then invite the poorest section of the villages—the landless labourers and land-owning cultivators with small holdings to reclaim the ravines as noted above and the land may be settled with these people on some nominal cost and revenue. The craze of land will induce these poor farmers to work hard. In reclaiming one *katha* (1/32 of an acre) of medium and small ravines a maximum of ten labour-days would be needed. An acre would need 320 labour-days. This at the present wage rate would be about Rs. 1,000 per acre. In the areas of large gullies the cost may be several times —say five times and more.

Now, if a moderate harvest of some crop, *e.g.* millet, maize, oilseeds or pulses or groundnut is raised on such an acre, a moderate yield of 5-8 quintals an acre would fetch an annual income of about Rs. 200-300. If two harvests are raised, it would be Rs. 400-600 a year and the labour cost on reclamation would be returned within a few years. But such great care and devotion that these newly reclaimed fields would need can come only from the land 'have-nots' on the guarantee that the land would be theirs. Caution may be taken in such a reclamation so that no agrarian unrest is unnecessarily created. If the original owners want to maintain their rights (provided they can do so under the changed land tenancy acts—particularly ceilings), they may be allowed to do so provided they undertake to reclaim such land within a stipulated time. As most of the ravine zones occur near some rivers, it may be possible to install water-pumping sets in the areas of reclamation where water will be so essential. Water flows through most of the streams during monsoon and post-monsoon period and pumping for the irrigation of at least one crop would be possible.

(*vi*) *Diversion Channels*: Diversion channels are essential in most reclamation schemes in gullied tracts because unless water is diverted, its flow into the ravine would damage the conservation measures whether they are grasses or plants or earthen or masonry or brush and wire dams. The diversion channel would take away the discharge of the catchment into a different area and the amount of water flow would diminish in the gullies which would cease to be damaged and would stop growing when the reclamation measures would gradually diminish their magnitude and would ultimately obliterate them.

In the areas marked by gully erosion, any unnatural inequality of surface starts off gully action. Thus, the depression. formed along roads and railways or canals or for brick-making or building sites, etc., by the earth excavated for embankment have initiated gully erosion. *e.g.* along the National Highway between Ranchi. and Jamshedpur or near the Medical College, Ranchi. The diversion channels, mentioned above, lest they should initiate the gully which they intend to check, should be grass-covered.

(*vii*) *Field bunding and embodiments*: This measure is being adopted in the districts of the worst ravines, *e.g.* Gwalior and Mathura (Fig. 14). Such surface is relatively level and occurs in the marginal areas of small and moderate gullies. Individual fields may be bunded, so that such reclamation measures pertain to an individual plot. On a large scale an embankment enclosing a greater area of several fields may be constructed to cut off the area from the rest of the ravine zone. Detailed conservation measures may then be applied to the area.

Conservation measures in the areas of marine erosion

At present conservation measures are in progress in the coastal stretches of the worst marine erosion, *i.e.* in Kerala.

(*i*) The method adopted is the construction of *groynes* or *walls of granite*. It is a costly method involving an expenditure of about Rs. 12 lakhs to Rs. 15 lakhs per mile of coast so protected. It is only the vital points, therefore, of more serious erosion affecting trunk road, railways and navigation canals, etc., that are being protected.

(*ii*) It is estimated that about half of the coastal stretch under marine erosion in Kerala would be protected by 1972[1]

[1] *The Statesman*, July 13, 1966.

(*iii*) The second conservation method is beach feeding or 'artificial nourishment' of the beach by dumping sand on the beach so that the latter may be stabilised. It would require such costly equipment as sand dumpers, earth-movers, and an exhaustive study of tides, winds and movement of coastal waters. It is regarded costlier than groynes.

Conservation measures against wind erosion

In a country of limited resources like India, where intensive agriculture is common, not much can be done towards checking wind erosion. As for example, it is difficult to think of tree shelter-belts because such growth reduces the production of the fields that fall in the immediate shadow zone of such vegetation. The loss thus entailed may be more than what could be compensated by checking wind erosion.

As mentioned earlier, it is the loose sandy and light loamy soils of the desert or semi-arid Gujarat or the newer alluvial belts of the Indo-Gangetic Plain that suffer most from wind erosion. The greatest safety against wind erosion could be provided by moisture in the soil. Unfortunately, the high wind period in the areas noted above coincides with that of the highest temperature, maximum evaporation, minimum atmospheric humidity and precipitation and lowest soil moisture except in the water-logged areas found in the irrigated zones. Again few conservation measures can prove effective against the violent duststorms that occur in the hot dry weather season in northern India.

(*i*) *Protection from stubble and weeds*: There appears, however, to be an unconscious action on the part of the Indian agriculturists that acts against wind erosion. The *rabi* crops are harvesed towards the end of March and early April when the hot and strong *loo* winds start. The stubble and grasses and weeds that accompanied the *rabi* harvest are left standing in the field. Spading and ploughing is generally done towards the end of the hot dry weather after the crops have been thrashed and winnowed, grains collected and straw stacked and the cultivator has taken some rest. By the time the tillage starts the high *loo* winds may have died down and in many parts a few showers might have moistened the soil.

(*ii*) *If tillage is done at right angles to the direction of prevailing winds,* it would not involve any additional cost and would go a long way to reduce wind erosion. In northern Indian Plain the direction of the prevailing wind during the hot weather when wind erosion occurs is from west and if the fields are ploughed north-south the furrows and ridges would reduce wind erosion. On sloping lands, however, where contour tillage is essential, it would be difficult to orient tillage at right angles to the wind.

(iii) Strip cultivation can be quite effective against wind erosion. We have such crops as sugarcane, cotton, which are cultivated side by side with such crops as wheat, barley, sweet potato, small millet, pulses and autumn paddy, etc., on the *khadar* areas of North Indian Plain. If the crops of the former category, *i.e.* sugarcane or cotton which keep standing in the field during the hot dry season are grown in strips parallel with those of the crops of the other category, wind erosion can be checked by the 'wind-break' effect of the standing crops.

Other refined methods, *e.g.* fences, burlap screens and other wind-breaks, grass cultivation and strip-cropping can hardly work in the Indian scene.

Conservation measure in the areas of gracial erosion : In the relatively sparsely populated Himalayan areas of glacial erosion, mainly by avalanches in the periglacial belt below the permanent snow line, conservation measures are difficult. The planting of trees or the construction of walls of timber logs[1] on sites which are particularly steep and amenable to avalanches are the measures adopted by the villagers for protecting their terraced fields and habitations.

[1] Pant, S. D., *The Social Economy of the Himalayans* (London, 1935).

INDEX